ASCE/SEI/SFPE 29-99

American Society of Civil Engineers
Society of Fire Protection Engineers

Standard Calculation Methods for Structural Fire Protection

This document uses both the International System of Units (SI) and customary units.

ABSTRACT

SEI/ASCE/SFPE Standard 29-99, *Standard Calculation Methods for Structural Fire Protection,* provides methods to calculate the fire resistance of selected structural members and barrier assemblies using structural steel, plain concrete, reinforced concrete, timber and wood, concrete masonry, and clay masonry. These methods are intended to provide architects, engineers, building officials, and others with calculation methods that will give the equivalent fire resistance that would have been achieved in the ASTM E119 standard fire test.

Library of Congress Cataloging-in-Publication Data

Standard calculation methods for structural fire protection /
  American Society of Civil Engineers, Society of Fire Protection
  Engineers.
    p. cm.
  Includes bibliographical references and index.
  ISBN 0-7844-0649-9
    1. Building, Fireproof—Standards—United States. 2. Numerical
analysis—Standards—United States.   I. American Society of
Civil Engineers. II. Society of Fire Protection Engineers.

TH1065.S684 2002
693.8'2—dc21

2002043757

STANDARDS

In April 1980, the Board of Direction approved ASCE Rules for Standards Committees to govern the writing and maintenance of standards developed by the Society. All such standards are developed by a consensus standards process managed by the Management Group F (MGF), Codes and Standards. The consensus process includes balloting by the balanced standards committee made up of Society members and nonmembers, balloting by the membership of ASCE as a whole, and balloting by the public. All standards are updated or reaffirmed by the same process at intervals not exceeding 5 years.

The following Standards have been issued:

ANSI/ASCE 1-82 N-725 Guideline for Design and Analysis of Nuclear Safety Related Earth Structures

ANSI/ASCE 2-91 Measurement of Oxygen Transfer in Clean Water

ANSI/ASCE 3-91 Standard for the Structural Design of Composite Slabs and ANSI/ASCE 9-91 Standard Practice for the Construction and Inspection of Composite Slabs

ASCE 4-98 Seismic Analysis of Safety-Related Nuclear Structures

Building Code Requirements for Masonry Structures (ACI 530-02/ASCE 5-02/TMS 402-02) and Specifications for Masonry Structures (ACI 530.1-02/ASCE 6-02/TMS 602-02)

SEI/ASCE 7-02 Minimum Design Loads for Buildings and Other Structures

SEI/ASCE 8-02 Standard Specification for the Design of Cold-Formed Stainless Steel Structural Members

ANSI/ASCE 9-91 listed with ASCE 3-91

ASCE 10-97 Design of Latticed Steel Transmission Structures

SEI/ASCE 11-99 Guideline for Structural Condition Assessment of Existing Buildings

ANSI/ASCE 12-91 Guideline for the Design of Urban Subsurface Drainage

ASCE 13-93 Standard Guidelines for Installation of Urban Subsurface Drainage

ASCE 14-93 Standard Guidelines for Operation and Maintenance of Urban Subsurface Drainage

ASCE 15-98 Standard Practice for Direct Design of Buried Precast Concrete Pipe Using Standard Installations (SIDD)

ASCE 16-95 Standard for Load and Resistance Factor Design (LRFD) of Engineered Wood Construction

ASCE 17-96 Air-Supported Structures

ASCE 18-96 Standard Guidelines for In-Process Oxygen Transfer Testing

ASCE 19-96 Structural Applications of Steel Cables for Buildings

ASCE 20-96 Standard Guidelines for the Design and Installation of Pile Foundations

ASCE 21-96 Automated People Mover Standards—Part 1

ASCE 21-98 Automated People Mover Standards—Part 2

ASCE 21-00 Automated People Mover Standards—Part 3

SEI/ASCE 23-97 Specification for Structural Steel Beams with Web Openings

SEI/ASCE 24-98 Flood Resistant Design and Construction

ASCE 25-97 Earthquake-Actuated Automatic Gas Shut-Off Devices

ASCE 26-97 Standard Practice for Design of Buried Precast Concrete Box Sections

ASCE 27-00 Standard Practice for Direct Design of Precast Concrete Pipe for Jacking in Trenchless Construction

ASCE 28-00 Standard Practice for Direct Design of Precast Concrete Box Sections for Jacking in Trenchless Construction

SEI/ASCE/SFPE 29-99 Standard Calculation Methods for Structural Fire Protection

SEI/ASCE 30-00 Guideline for Condition Assessment of the Building Envelope

SEI/ASCE 32-01 Design and Construction of Frost-Protected Shallow Foundations

EWRI/ASCE 33-01 Comprehensive Transboundary International Water Quality Management Agreement

EWRI/ASCE 34-01 Standard Guidelines for Artificial Recharge of Ground Water

EWRI/ASCE 35-01 Guidelines for Quality Assurance of Installed Fine-Pore Aeration Equipment

CI/ASCE 36-01 Standard Construction Guidelines for Microtunneling

SEI/ASCE 37-02 Design Loads on Structures During Construction

CI/ASCE 38-02 Standard Guideline for the Collection and Depiction of Existing Subsurface Utility Data

EWRI/ASCE 39-03 Standard Practice of the Design and Operation of Hail Suppression Projects

FOREWORD

In April 1995, the Board of Direction approved the revision to the ASCE Rules for Standards Committees to govern the writing and maintenance of Standards developed by the Society. All such Standards are developed by a consensus standards process managed by the ASCE Codes and Standards Activities Committee (CSAC). The consensus process includes balloting by a balanced standards committee made up of Society members and nonmembers, balloting by the membership of ASCE as a whole, and balloting by the public. All Standards are updated or reaffirmed by the same process at intervals not exceeding 5 years.

The material presented in this Standard has been prepared in accordance with recognized engineering principles. This Standard should not be used without first securing competent advice with respect to its suitability for any given application. The publication of the material contained herein is not intended as a representation or warranty on the part of the American Society of Civil Engineers, or of any other person named herein, that this information is suitable for any general or particular use or promises freedom from infringement of any patent or patents. Anyone making use of this information assumes all liability from such use.

ACKNOWLEDGMENTS

The Structural Engineering Institute (SEI) of the American Society of Civil Engineers (ASCE) acknowledges the devoted efforts of the Structural Design for Fire Conditions Standards Committee of the Codes and Standards Activities Division. This group comprises individuals from many backgrounds including consulting engineering, research, construction industry, education, government, design, and private practice.

The development of this Standard was a joint effort between SEI and the Society of Fire Protection Engineers (SFPE). Although developed through ASCE's consensus process, SFPE contributed greatly to the development of this Standard.

This Standard was prepared through the consensus standards process by balloting in compliance with procedures of ASCE's Codes and Standards Activities Committee. Those individuals who serve on the Structural Design for Fire Conditions Standards Committee are:

James P. Barris, Chair
Robert W. Fitzgerald, Vice-Chair
Jesse J. Beitel
Kenneth E. Bland
Richard W. Bletzacker
Joseph A. Bohinsky
Delbert F. Boring
Richard W. Bukowski
Susan Dowty
Robert H. Dutson
Arobindo Dutt
Joseph M. Englot
John A. Frauenhoffer
Daniel F. Gemery
Ronald R. Gerace

Clayford T. Gimm
Ram A. Goel
Peter J. Gore Willse
Alfred G. Handy
Mark B. Hogan
Craig A. Holmes
John W. Home
Robert C. Jackson
Jack Jones
Donal H. Landis
Tiam T. Lie
Sadek W. Mansour
Edward F. Martella
John H. Matthys
Robert R. McCluer

Joseph J. Messersmith, Jr.
James A. Milke, Vice-Chair
Arthur J. Mullkoff
Mark A. Nunn
Walter J. Prebis
William D. Rome
Joseph E. Saliba
Erwin L. Schaffer
Kenneth M. Schoonover
Paul D. Sullivan
Jan S. Teraszkiewicz
Phillip C. Terry
Michael G. Verlaque
Roger H. Wildt
George William

CONTENTS

CONTENTS

Figures

CONTENTS

Tables

Standard Calculation Methods for Structural Fire Protection

1. STANDARD CALCULATION METHODS FOR STRUCTURAL FIRE PROTECTION

1.1 General

Building codes specify the fire resistance required for structural members and barriers in identified occupancies and classifications of construction. The fire endurance is based on the test results of the American Society for Testing and Materials (ASTM) test designation E119, *Standard Test Methods for Fire Tests of Buildings Construction and Materials.*

As an alternative to selection of tested assemblies, this Standard provides methods to calculate the fire resistance of selected structural member and barrier assemblies using structural steel, plain concrete, reinforced concrete, timber and wood, concrete masonry, and clay masonry. These methods are intended to provide architects, engineers, building officials, and others with calculation methods that will give the equivalent fire resistance that would have been achieved in the ASTM E119 standard fire test.

1.2 Scope
1.2.1

The calculation methods provided in the document are intended to produce fire resistance rating times that are equivalent to the results obtained from the standard fire test, ASTM E119, *Standard Test Methods for Fire Tests of Building Construction and Materials.* The calculation methods of this Standard are for use as an alternative to the laboratory test results.

1.2.2

These calculation methods are applicable only to structural steel, plain concrete, reinforced concrete, timber and wood, concrete masonry, and clay masonry. Limitations of applicability are identified in the individual chapters that describe the methods for each of the materials that comprise this Standard.

1.3 Purpose and Use
1.3.1

While the fire resistance ratings calculated by the procedures specified in this Standard are equivalent substitutes for the results obtained by the ASTM E119 standard fire test, they do not necessarily describe the performance for natural fires having time-temperature relationships different from ASTM E119.

1.3.2

The fire resistance results obtained by calculation methods are for use in building fire evaluations or for building code applications. It is the responsibility of the user of this Standard to establish appropriate technical or regulatory use for the results.

1.3.3

The procedures for calculating the fire resistance ratings for structural members or assemblies for the different structural materials are organized under the following chapters:

Chapter 2. Standard Methods for Determining the Fire Resistance of Plain and Reinforced Concrete Construction

Chapter 3. Standard Methods for Determining the Fire Resistance of Timber and Wood Structural Elements

Chapter 4. Standard Calculation Methods for Determining the Fire Resistance of Masonry

Chapter 5. Standard Methods for Determining the Fire Resistance of Structural Steel Construction

1.4 Referenced Standards
1.4.1 American Concrete Institute (ACI)

ACI 318-95 Building Code Requirements for Structural Concrete

ACI 530-95/ASCE 5-95/TMS 402-95 Building Code Requirements for Masonry Structures

1.4.2 American Society for Testing and Materials (ASTM)

ASTM C33-93	Standard Specification for Concrete Aggregates
ASTM C67-94	Standard Methods of Sampling and Testing Brick and Structural Clay Tile
ASTM C140-95a	Standard Methods of Sampling and Testing Concrete Masonry Units
ASTM C331-94	Standard Specification for Lightweight Aggregates for Concrete Masonry Units
ASTM C332-87	Standard Specification for Lightweight Aggregates for Insulating Concrete (Reapproved 1991)
ASTM C612-93	Standard Specification for Mineral Fiber Block and Board Thermal Insulation

1

ASTM C726-93   Standard Specification for Mineral Fiber Roof Insulation Board

ASTM C796-87a   Standard Test Method for Foaming for Use in Producing Cellular Concrete Using Preformed Foam (Reapproved 1992)

ASTM E119-95a   Standard Test Methods for Fire Tests of Building Construction and Materials

1.4.3 American Forest & Paper Association (AF&PA)

NDS-91   National Design Specification for Wood Construction

1.4.4 American Institute of Steel Construction (AISC)

LRFD-94   Load and Resistance Factor Design Specification for Structural Steel Buildings

1.5 Definitions

Approved: Acceptable to the authority having jurisdiction.

Authority having jurisdiction: The organization, political subdivision, office, or individual charged with the responsibility of administering and enforcing the provisions of this Standard.

2. STANDARD METHODS FOR DETERMINING THE FIRE RESISTANCE OF PLAIN AND REINFORCED CONCRETE CONSTRUCTION

2.1 Scope

This Section describes procedures for determining the fire resistance rating of plain concrete walls and reinforced concrete walls, floors, roofs, beams, and columns by calculation. These provisions shall apply to concrete made with cementitious materials, aggregates, and admixtures permitted by ACI 318, except that the specified compressive strength of concrete, f'_c, used in the design shall not exceed 10,000 psi (69 MPa). Except where the provisions of this Chapter are more stringent, concrete members shall comply with ACI 318.

These provisions shall apply to concrete slabs cast on stay-in-place non-composite steel forms where the slab is reinforced to carry all superimposed loads and the dead load of the slab. These provisions shall not apply to concrete slabs cast on stay-in-place non-composite steel forms where the form is designed to carry the dead load of the slab or to composite slabs where the steel form serves as the positive moment reinforcement.

2.2 Definitions

Words and terms used in this Chapter shall have the following meanings:

Carbonate aggregate concrete: Concrete made with aggregate consisting mainly of calcium or magnesium carbonate (e.g., limestone or dolomite).

Cellular concrete: A nonstructural lightweight insulating concrete made by mixing a preformed foam with portland cement slurry and having a dry unit weight of approximately 30 lbs per cu ft (487 kg/m^3) determined in accordance with ASTM C796.

Ceramic fiber blanket: A mineral wool insulating material made of alumina-silica fibers and having a density of 4 to 8 lbs per cu ft (65 to 130 kg/m^3).

Glass fiberboard: Fibrous glass roof insulation board complying with ASTM C612.

Lightweight aggregate concrete: Concrete made with aggregates of expanded clay, shale, slag, or slate or sintered fly ash, and having a dry unit weight of 85 to 115 lbs per cu ft (1,362 to 1,842 kg/m^3).

Mineral board: Mineral fiber roof insulation board complying with ASTM C726.

Perlite concrete: A nonstructural lightweight insulating concrete having a dry unit weight of approximately 30 lbs per cu ft (481 kg/m^3) made by mixing perlite concrete aggregate complying with ASTM C332 with portland cement slurry.

Plain concrete: Concrete that does not conform to the requirements for reinforced concrete.

Reinforced concrete: Concrete reinforced with no less than the minimum amount of steel required by ACI 318, prestressed or non-prestressed, and designed on the basis that the two materials act together in resisting forces.

Sand-lightweight aggregate concrete: Concrete made with a combination of expanded clay, shale, slag or slate or sintered fly ash, and natural sand, and having a dry unit weight between 105 and 120 lbs per cu ft (1,682 and 1,922 kg/m^3).

Siliceous aggregate concrete: Concrete made with aggregates consisting mainly of silica or compounds other than calcium or magnesium carbonate.

Vermiculite concrete: A nonstructural lightweight insulating concrete having a dry unit weight of approximately 30 lbs/cu ft (481 kg/m^3) made by mixing vermiculite concrete aggregate complying with ASTM C332 with portland cement slurry.

2.3 Concrete Walls

The minimum equivalent thickness of different types of plain or reinforced concrete bearing or nonbearing walls required to provide fire resistance ratings of 1 to 4 hours shall be not less than that indicated in Table 2-1. For solid walls with flat surfaces, the equivalent thickness shall be the actual thickness. The equivalent thickness of hollow-core walls or of walls with surfaces that are not flat shall be determined in accordance with Sections 2.3.1 through 2.3.3.

2.3.1 Hollow-Core Panel Walls

For walls constructed with precast hollow-core panels with constant core cross section throughout their length, the equivalent thickness shall be the net cross-sectional area divided by the panel width. Where all of the core spaces are filled with grout or loose fill material, such as perlite, vermiculite, sand or expanded clay, shale, slag, or slate, the fire resistance rating of the wall shall be considered the same as that of a solid wall of the same type of concrete.

2.3.2 Flanged Wall Panels

For walls constructed with flanged wall panels where the flanges taper, the equivalent thickness shall be determined at a distance of two times the minimum thickness or 6 in. (152 mm), whichever is less, from the point of minimum thickness.

2.3.3 Ribbed or Undulating Panels

The equivalent thickness, T_e, of panels with ribbed or undulating surfaces shall be determined as follows:

a. Where the spacing of ribs or undulations is equal to or greater than four times the minimum thickness, the equivalent thickness is the minimum thickness.
b. Where the spacing of ribs or undulations is equal to or less than two times the minimum thickness, the equivalent thickness is calculated by dividing the net cross-sectional area by the panel width. The maximum thickness used to calculate the net cross-sectional area shall not exceed two times the minimum thickness.
c. Where the spacing of ribs or undulations exceeds two times the minimum thickness but is less than

four times the minimum thickness, the equivalent thickness is calculated from the following formula:

$$T_e = t + [(4t/s) - 1](t_e - t) \qquad (2.1)$$

where

s = spacing of ribs or undulations
t = minimum thickness
t_e = equivalent thickness calculated in accordance with item b, above

2.3.4 Multiple-Wythe Walls

For walls consisting of two or more wythes of different types of concrete, the fire resistance rating shall be determined in accordance with the graphical or numerical solution in Sections 2.3.4.1 and 2.3.4.2, respectively.

2.3.4.1 Graphical Solution: For walls consisting of two wythes of different types of concrete, the fire resistance rating shall be determined from Figure 2-1. The fire resistance rating shall be the lower of the two ratings determined by assuming that each side of the wall is the fire-exposed side.

2.3.4.2 Numerical Solution: For walls consisting of two or more wythes of different types of concrete, or one or more wythes of concrete and one or more wythes of masonry, the fire resistance rating shall be determined from the formula:

$$R = (R_1^{0.59} + R_2^{0.59} + ... + R_n^{0.59})^{1.7} \qquad (2.2)$$

where

R = fire resistance rating of assembly, minutes
R_1, R_2, R_n = fire resistance rating of individual wythes, minutes

TABLE 2-1. Fire Resistance of Concrete Walls, Floors, and Roofs

Concrete Aggregate Type	Minimum Equivalent Thickness for Fire Resistance Rating (hrs)									
	1 hr		1.5 hr		2 hr		3 hr		4 hr	
	in.	mm	in.	mm	in.	mm	in.	mm	in.	mm
Siliceous	3.5	89	4.3	109	5.0	127	6.2	157	7.0	178
Carbonate	3.2	81	4.0	102	4.6	117	5.7	145	6.6	168
Sand-lightweight	2.7	69	3.3	84	3.8	97	4.6	117	5.4	137
Lightweight	2.5	64	3.1	79	3.6	91	4.4	112	5.1	130

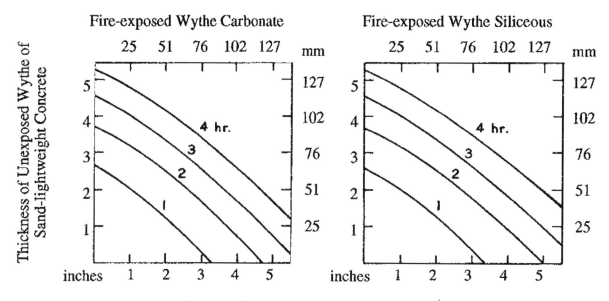

THICKNESS OF FIRE-EXPOSED WYTHE IN INCHES

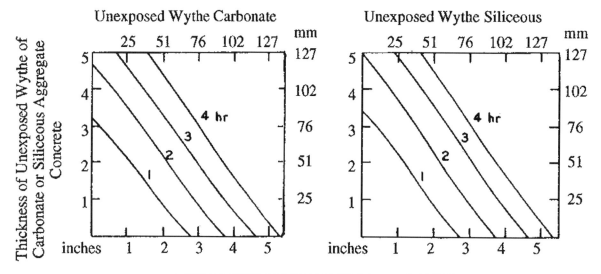

THICKNESS OF FIRE-EXPOSED WYTHE OF
SAND-LIGHTWEIGHT CONCRETE

FIGURE 2-1. Fire Resistance Ratings of Two-Wythe Concrete Walls (From Abrams, M.S. and A.H. Gusta-ferro. *Fire Endurance of Two-Course Floors and Roofs,* **Portland Cement Association Research and Development Bulletin RD048, 1968. Used with permission.)**

Values of R_n^{059} for individual wythes of concrete for use in the formula are obtained from Figure 2-2. Values for R_n for masonry are obtained from Chapter 4.

2.3.4.2.1 Sandwich Panels: The fire resistance rating of precast concrete wall panels consisting of a layer of foam plastic sandwiched between two wythes of concrete shall be determined by use of Equation 2.2. The R_n^{059} value for 1 in. (25 mm) or thicker foam plastic for

use in the equation is 2.5. The foam plastic shall be protected on both sides by not less than 1 in. (25 mm) of concrete. Foam plastic with a total thickness of less than 1 in. (25 mm) shall be disregarded.

2.3.4.2.2 Air Spaces: The fire resistance rating of concrete walls incorporating an air space between two wythes of concrete shall be determined by use of Equation 2.2. The R_n^{059} value for one .5-in.-wide

(13 mm) to 3.5 in. wide (89 mm) air space is 3.3. The R_n^{059} value for two .5-in.-wide (13 mm) to 3.5-in.-wide (89 mm) air spaces is 6.7.

2.3.5 Joints Between Precast Concrete Wall Panels

Joints between precast concrete wall panels required to be insulated by Section 2.3.5.1 shall be insulated in accordance with Section 2.3.5.2.

2.3.5.1 Joints Required to Be Insulated: Where openings are not permitted or where openings are required to be protected, the provisions of Section 2.3.5.2 shall be used to determine the required thickness of joint insulation.

Joints between concrete wall panels that are not insulated as required by Section 2.3.5.2 shall be considered unprotected openings. Uninsulated joints in exterior walls shall be included with other openings in determining the percentage of unprotected openings permitted by building code provisions. Insulated joints shall not be considered openings for purposes of determining compliance with the allowable percentage of openings.

2.3.5.2 Thickness of Insulation: The thickness of ceramic fiber blanket insulation required to insulate 3/8-in. (10 mm) wide and 1-in. (25 mm) wide joints between concrete wall panels to maintain fire resistance ratings

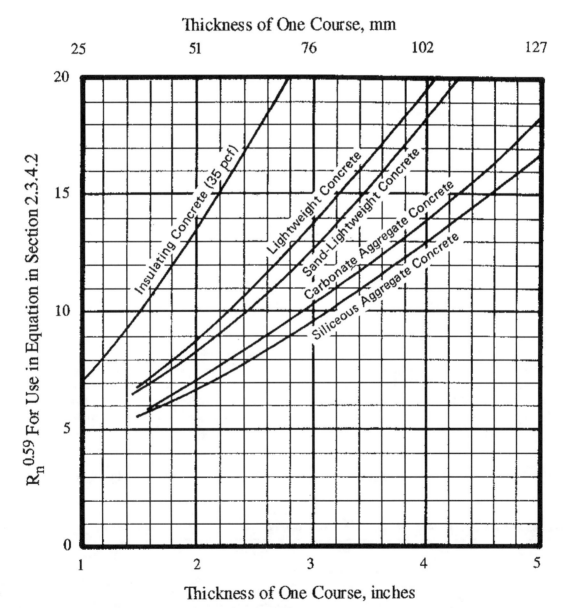

FIGURE 2-2. Values of $R_n^{0.59}$ for Different Types of Concrete

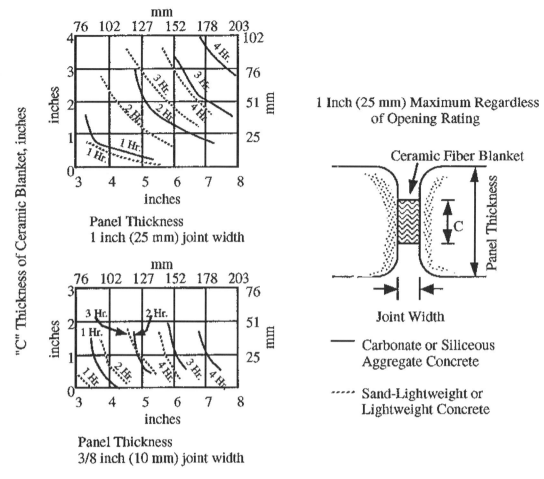

FIGURE 2-3. Ceramic Fiber Joint Protection

of 1 hour to 4 hours shall be in accordance with Figure 2-3. For joint widths between 3/8 in. (10 mm) and 1 in. (25 mm), the thickness of insulation shall be determined by direct interpolation. Other joint treatments shall not be used unless they are determined to maintain the required fire resistance and are approved.

2.3.6 Walls with Gypsum Wallboard or Plaster Finishes

The fire resistance rating of cast-in-place or precast concrete walls with finishes of gypsum wallboard or plaster applied to one or both sides of the wall shall be determined in accordance with this Section.

TABLE 2-2. Multiplying Factor for Finishes on Non–Fire-Exposed Side of Concrete Wall

Type of Finish Applied to Wall	Type of Aggregate Used in Concrete		
	Siliceous or Carbonate	Sand-Lightweight	Lightweight
Portland Cement-Sand Plaster	1.00	0.75	0.75
Gypsum-Sand Plaster	1.25	1.00	1.00
Gypsum-Vermiculite or Perlite Plaster	1.75	1.50	1.25
Gypsum Wallboard	3.00	2.25	2.25

2.3.6.1 Calculation for Non–Fire-Exposed Side:
Where the finish of gypsum wallboard or plaster is applied to the non–fire-exposed side of the wall, the fire resistance rating of the entire assembly shall be determined as follows. The thickness of the finish shall be adjusted by multiplying the actual thickness of the finish by the applicable factor from Table 2-2 based on the type of aggregate in the concrete. The adjusted finish thickness shall be added to the actual thickness or equivalent thickness of concrete and the fire resistance rating of the concrete, including finish, determined from Table 2-1, Figure 2-1, or Figure 2-2.

2.3.6.2 Calculation for Fire-Exposed Side: Where the finish of gypsum wallboard or plaster is applied to the fire-exposed side of the wall, the fire resistance rating of the entire assembly shall be determined as follows. The time assigned to the finish by Table 2-3 shall be added to the fire resistance rating of the concrete wall determined from Table 2-1, Figure 2-1, or Figure 2-2, or to the rating determined in accordance with Section 2.3.6.1 for the concrete and finish on the non–fire-exposed side.

2.3.6.3 Assume Each Side of Wall Is Fire-Exposed Side: Where a wall is required to be fire resistance rated from both sides and has no finish on one side or has different types and/or thicknesses of finish on each side, the calculation procedures of Sections 2.3.6.1 and 2.3.6.2 shall be performed twice (i.e., assume that each side of the wall is the fire-exposed side). The fire resistance rating of the wall, including finishes, shall not exceed the lower of the two values calculated.

2.3.6.4 Minimum Rating Provided by Concrete: Where finishes applied to one or both sides of a concrete wall contribute to the fire resistance rating, the concrete alone shall provide not less than one-half of the total required fire resistance rating. The contribution to fire resistance of the finish on the non–fire-exposed side of a load-bearing wall shall not exceed 0.5 times the contribution of the concrete alone.

2.3.6.5 Installation of Finishes: Finishes on concrete walls that contribute to the total required fire resistance rating shall comply with the installation requirements of Sections 2.3.6.5.1 and 2.3.6.5.2 and applicable building code provisions.

2.3.6.5.1 Furring Members: Gypsum wallboard and gypsum lath shall be secured to wood or steel furring members spaced not more than 16 in. (406 mm) on center.

TABLE 2-3. Time Assigned to Finish Materials on Fire-Exposed Side of Concrete Wall

Finish Description	Time (min)
Gypsum Wallboard	
3/8 in. (10 mm)	10
1/2 in. (13 mm)	15
5/8 in. (16 mm)	20
2 layers of 3/8 in. (20 mm)	25
1 layer of 3/8 in. (10 mm) and 1 layer of 1/2 in. (13 mm)	35
2 layers of 1/2 in. (25 mm)	40
Type X Gypsum Wallboard	
1/2 in. (13 mm)	25
5/8 in. (16 mm)	40
Portland Cement-Sand Plaster Applied Directly to Concrete	*
Portland Cement-Sand Plaster on Metal Lath	
3/4 in. (19 mm)	20
7/8 in. (22 mm)	25
1 in. (25 mm)	30
Gypsum Sand Plaster on 3/8 in. (10 mm) Gypsum Lath	
1/2 in. (13 mm)	35
5/8 in. (16 mm)	40
3/4 in. (19 mm)	50
Gypsum Sand Plaster on Metal Lath	
3/4 in. (19 mm)	50
7/8 in. (22 mm)	60
1 in. (25 mm)	80

* The actual thickness of portland cement-sand plaster shall be included in the determination of the equivalent thickness of the concrete for use in Table 2-1 only when it is 5/8-in. (16 mm) thick or less.

2.3.6.5.2 Gypsum Wallboard Orientation: Gypsum wallboard shall be installed with the long dimension parallel to furring members and shall have all horizontal and vertical joints supported and finished.

Exception: 5/8-in. (16 mm) thick Type X gypsum wallboard is not required to be installed with the long dimension parallel to the framing members or have the horizontal joints supported.

2.4 Concrete Floor and Roof Slabs

The minimum equivalent thickness of different types of concrete floor and roof slabs required to provide fire resistance ratings of 1 to 4 hours shall be not less than the thickness indicated in Table 2-1. For solid slabs with flat surfaces, the equivalent thickness shall be the same as the actual thickness.

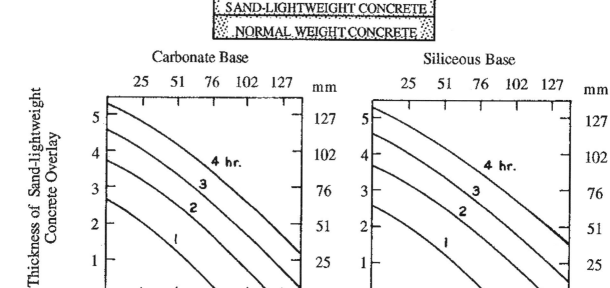

THICKNESS OF NORMAL WEIGHT CONCRETE BASE SLAB

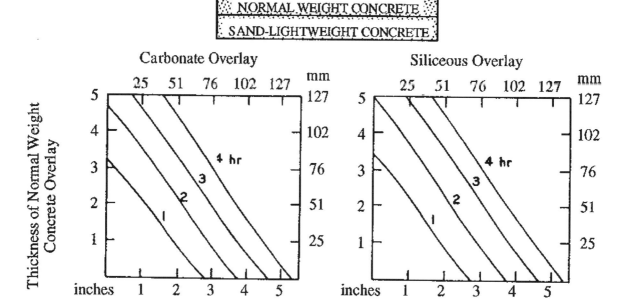

THICKNESS OF SAND-LIGHTWEIGHT CONCRETE BASE SLAB

FIGURE 2-4. Fire Resistance of Two-Course Concrete Floors and Roofs (From Abrams, M.S. and A.H. Gustaferro. *Fire Endurance of Two-Course Floors and Roofs,* **Portland Cement Association Research and Development Bulletin RD048, 1968. Used with permission.)**

2.4.1 Slabs with Other Than Flat Surfaces

For floors and roofs constructed with hollow-core panels, flanged members, or with ribbed or undulating surfaces, the minimum thickness or equivalent thickness shall be determined as required for hollow core panel walls, flanged wall panels and ribbed or undulating panels. See Sections 2.3.1, 2.3.2, and 2.3.3.

2.4.2 Joints in Precast Slabs

Joints between adjacent precast concrete slabs shall be ignored when calculating the slab thickness provided a concrete topping not less than 1 in. (25 mm) thick is used. Where a concrete topping is not used, joints shall be grouted to a depth of at least one-third the slab thickness at the joint, but not less than 1 in. (25 mm); or the fire resistance rating of the floor or roof shall be maintained by other approved methods.

2.4.3 Two-Course Floors and Roofs

The fire resistance rating of two-course floors and roofs consisting of a base slab of concrete with a topping (overlay) of concrete with a different type of aggregate is indicated in Figure 2-4. If the base slab of concrete is covered with a topping (overlay) of terrazzo or gypsum wallboard, the thickness of terrazzo or gypsum wallboard shall be converted to an equivalent thickness of concrete by multiplying the actual thickness by the appropriate factor listed in Table 2-4. This equivalent concrete thickness shall be added to the base slab thickness and the total thickness used to determine the fire resistance of the slab, including topping, from Table 2-1.

2.4.4 Insulated Roofs

The fire resistance rating of roofs consisting of a base slab of concrete with a topping (overlay) of cellular, perlite, or vermiculite concrete or insulating boards and built-up roof shall be determined from Figure 2-5 (a) or (b). Where a three-ply built-up roof is installed over a lightweight insulating concrete topping, 10 minutes shall be permitted to be added to the fire resistance rating calculated from Figure 2-5 (a).

2.5 Concrete Cover over Reinforcement

Minimum concrete cover over positive moment reinforcement for floor and roof slabs and beams shall be determined by Sections 2.5.1 through 2.5.3. Concrete cover shall not be less than required by ACI 318. For purposes of determining minimum concrete cover, slabs and beams shall be classified as restrained or unrestrained in accordance with Appendix A.

2.5.1 Cover for Slab Reinforcement

The minimum thickness of concrete cover to positive moment non-prestressed and prestressed reinforcement (bottom steel) for different types of concrete floor and roof slabs for fire resistance ratings of 1 to 4 hours shall be not less than the thickness indicated in Table 2-5. For floor or roof slabs consisting of two or more courses of different types of concrete, the cover requirements shall be based on the type of concrete used for the base slab, provided the base slab is not less than 1 in. (25 mm) thick. Table 2-5 shall apply to one-way or two-way cast-in-place or precast solid or hollow-core slabs with flat undersurfaces.

2.5.2 Cover for Non-Prestressed Reinforcement in Beams

The minimum thickness of concrete cover to positive moment non-prestressed reinforcement (bottom steel) for restrained and unrestrained beams of different widths for fire resistance ratings of 1 to 4 hours for all types of concrete shall not be less than the thickness indicated in Table 2-6. Values in Table 2-6 for restrained beams apply to beams spaced more than 4 ft (1,219 mm) on center. For restrained beams and joists spaced 4 ft (1,219 mm) or less on center, 3/4-in. (19 mm) cover is adequate for fire resistance ratings of 4 hours or less regardless of the width of the beam. Cover for intermediate beam widths shall be determined by direct interpolation.

2.5.2.1 Calculating Cover: The concrete cover for an individual bar is the minimum thickness of concrete between the surface of the bar and the fire-exposed surface of the beam. For beams in which several bars are used, the cover, for the purposes of Table 2-6, is the average of the minimum cover of the individual bars. For corner bars (i.e., bars equal distance from the bottom and side), the minimum cover used in the calculation shall be one-half the actual value. The actual

TABLE 2-4. Multiplying Factors for Equivalent Thickness

	Base Slab Concrete Type	
Top Course Material	Siliceous or Carbonate	Sand-Lightweight or Lightweight
Gypsum Wallboard*	3	2.25
Terrazzo	1	0.75

* Applies only to roofs.

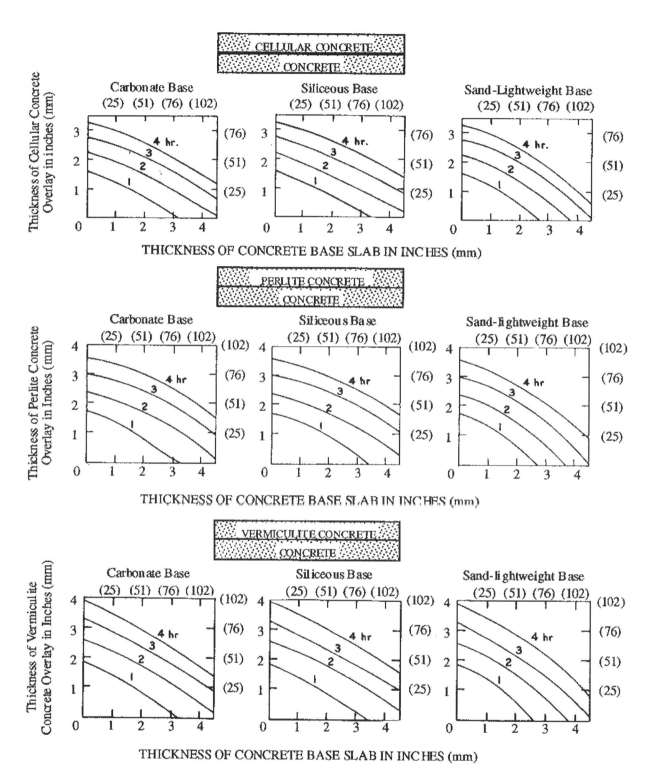

FIGURE 2-5 (a). Fire Resistance of Concrete Roofs with Overlays of Insulating Concrete (From Abrams, M.S. and A.H. Gustaferro. *Fire Resistance of Lightweight Insulating Concretes,* Research and Development Bulletin RD004, Portland Cement Association, 1970. Used with permission.)

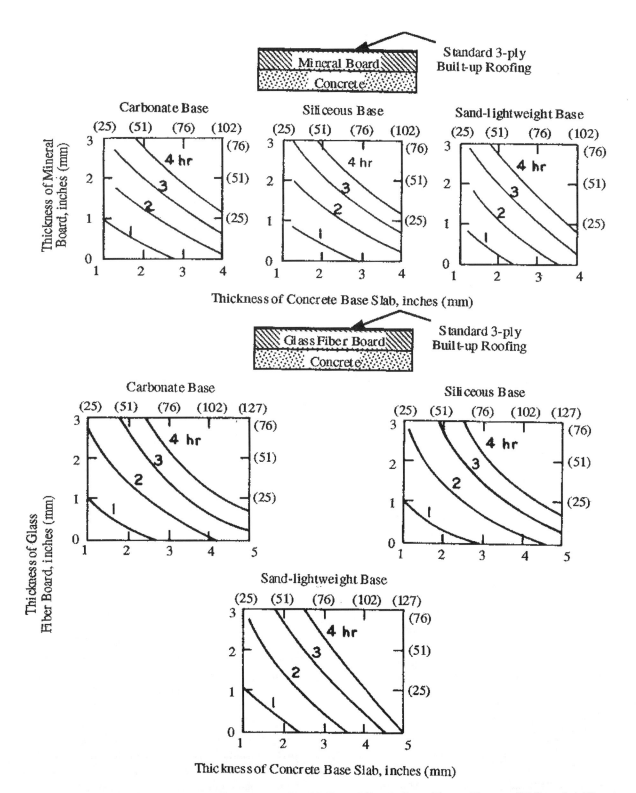

FIGURE 2-5 (b). Fire Resistance of Concrete Roofs with Board Insulations (From Abrams, M.S. and A.H. Gustaferro. *Fire Endurance of Two-Course Floors and Roofs.* Portland Cement Association Research and Development Bulletin RD048, 1968. Used with permission.)

11

TABLE 2-5. Minimum Cover for Non-Prestressed and Prestressed Reinforcement in Concrete Floor and Roof Slabs

Concrete Aggregate Type	Restrained		Thickness of Cover for Fire Resistance Rating									
			Unrestrained									
	4 hr or less		1 hr		1.5 hr		2 hr		3 hr		4 hr	
Minimum cover for non-prestressed reinforcement in concrete floor or roof slabs												
	in.	mm	in.	mm	in.	mm	in.	mm	in.	mm	in.	mm
Siliceous	3/4	19	3/4	19	3/4	19	1	25	1 1/4	32	1 5/8	41
Carbonate	3/4	19	3/4	19	3/4	19	3/4	19	1 1/4	32	1 1/4	32
Sand-lightweight	3/4	19	3/4	19	3/4	19	3/4	19	1 1/4	32	1 1/4	32
Lightweight	3/4	19	3/4	19	3/4	19	3/4	19	1 1/4	32	1 1/4	32
Minimum cover for prestressed reinforcement in concrete floor or roof slabs												
Siliceous	3/4	19	1 1/8	29	1 1/2	38	1 3/4	44	2 3/8	60	2 3/4	70
Carbonate	3/4	19	1	25	1 3/8	35	1 5/8	41	2 1/8	54	2 1/4	57
Sand-lightweight	3/4	19	1	25	1 3/8	35	1 1/2	38	2	51	2 1/4	57
Lightweight	3/4	19	1	25	1 3/8	35	1 1/2	38	2	51	2 1/4	57

TABLE 2-6. Minimum Cover for Non-Prestressed Reinforcement in Concrete Beams

Restrained or Unrestrained	Beam Width		Thickness of Cover for Fire Resistance Rating									
			1 hr		1.5 hr		2 hr		3 hr		4 hr	
	in.	mm	in.	mm	in.	mm	in.	mm	in.	mm	in.	mm
Restrained	5	127	3/4	19	3/4	19	3/4	19	1	25	1 1/4	32
Restrained	7	178	3/4	19	3/4	19	3/4	19	3/4	19	3/4	19
Restrained	≥10	≥254	3/4	19	3/4	19	3/4	19	3/4	19	3/4	19
Unrestrained	5	127	3/4	19	1	25	1 1/4	32	—	—	—	—
Unrestrained	7	178	3/4	19	3/4	19	3/4	19	1 3/4	44	3	76
Unrestrained	≥10	≥254	3/4	19	3/4	19	3/4	19	3/4	19	1 3/4	44

cover for an individual bar shall not be less than one-half the value shown in Table 2-6 or 3/4-in. (19 mm), whichever is greater.

2.5.3 Cover for Prestressed Reinforcement in Beams

The minimum thickness of concrete cover to positive moment prestressed reinforcement (bottom steel) for restrained and unrestrained beams of different types of concrete for fire resistance ratings of 1 to 4 hours shall be not less than the thickness indicated in Table 2-7 (a) and (b). Table 2-7 (a) shall apply to beams of all widths, provided the beam cross-sectional area is not less than 40 sq in. (25,806 mm²). Table 2-7 (b) shall apply to beam widths equal to or greater than 8 in. (203 mm). As an alternative, for beams with cross-sectional areas equal to or greater

than 40 sq in. (25,806 mm²) and widths equal to or greater than 8 in. (203 mm), the minimum concrete cover specified from the two tables shall be used. Values in Table 2-7 (a), and (b) for restrained beams shall apply to beams spaced more than 4 ft (1,219 mm) on center. For restrained joists spaced 4 ft (1,219 mm) or less on center, 3/4-in. (19 mm) cover shall be considered adequate for fire resistance ratings of 4 hours or less regardless of the cross-sectional area or width. Cover for intermediate beam widths of Table 2-7 (b) shall be determined by direct interpolation. When computing the cross-sectional area of beams cast monolithically with the supported slab for use in Table 2-7 (a) the cross-sectional area of the portion of the slab having a width not exceeding three times the average width of the beam shall be permit-

TABLE 2-7 (a). Minimum Cover for Prestressed Reinforcement in Concrete Beams 40 sq in. (1016 sq mm) or Greater in Area Regardless of Beam Widths

| Concrete Aggregate Type | Cross-Sectional Area | | Thickness of Cover for Fire Resistance Rating | | | | | | | | | |
| | | | 1 hr | | 1.5 hr | | 2 hr | | 3 hr | | 4 hr | |
	in.²	mm² × 10³	in.	mm	in.	mm	in.	mm	in.	mm	in.	mm
RESTRAINED												
All	40–150	26–97	1 1/2	38	1 1/2	38	2	51	2 1/2	64	—	—
Carbonate or Siliceous	>150–300	>97–194	1 1/2	38	1 1/2	38	1 1/2	38	1 3/4	44	2 1/2	64
Carbonate or Siliceous	>300	>194	1 1/2	38	1 1/2	38	1 1/2	38	1 1/2	38	2	51
Lightweight or Sand-lightweight	>150	>97	1 1/2	38	1 1/2	38	1 1/2	38	1 1/2	38	2	51
UNRESTRAINED												
All	40–150	26–97	2	51	2 1/2	64	—	—	—	—	—	—
Carbonate or Siliceous	>150–300	>97–194	1 1/2	38	1 3/4	44	2 1/2	64	—	—	—	—
Carbonate or Siliceous	>300	>194	1 1/2	38	1 1/2	38	2	51	3	76	4	102
Lightweight or Sand-lightweight	>150	>97	1 1/2	38	1 1/2	38	2	51	3	76	4	102

TABLE 2-7 (b). Minimum Cover for Prestressed Reinforcement in Concrete Beams 8 in. (203 mm) or Greater in Width

| Concrete Aggregate Type | Beam Width | | Thickness of Cover for Fire Resistance Rating | | | | | | | | | |
| | | | 1 hr | | 1.5 hr | | 2 hr | | 3 hr | | 4 hr | |
	in.	mm	in.	mm	in.	mm	in.	mm	in.	mm	in.	mm
RESTRAINED												
Carbonate or Siliceous	8	203	1 1/2	38	1 1/2	38	1 1/2	38	1 3/4	44	2 1/2	64
Carbonate or Siliceous	≥12	≥305	1 1/2	38	1 1/2	38	1 1/2	38	1 1/2	38	1 7/8	47
Sand-lightweight	8	203	1 1/2	38	1 1/2	38	1 1/2	38	1 1/2	38	2	51
Sand-lightweight	≥12	≥305	1 1/2	38	1 1/2	38	1 1/2	38	1 1/2	38	1 5/8	41
UNRESTRAINED												
Carbonate or Siliceous	8	203	1 1/2	38	1 3/4	44	2 1/2	64	5*	127	—	—
Carbonate or Siliceous	≥12	≥305	1 1/2	38	1 1/2	38	1 7/8	47	2 1/2	64	3	76
Sand-lightweight	8	203	1 1/2	38	1 1/2	38	2	51	3 1/4	95	—	—
Sand-lightweight	≥12	≥305	1 1/2	38	1 1/2	38	1 5/8	41	2	51	2 1/2	64

* Not practical for 8-in. (203-mm) wide beam, but shown for purposes of interpretation.

ted to be included with the cross-sectional area of the beam. Where the thickness of concrete cover exceeds 2.5 in. (64 mm), stirrups or hoops with a cover of 1 in. (25 mm) and spaced not to exceed the depth of the beam shall be provided. The minimum cover for non-prestressed positive moment reinforcement in prestressed concrete beams shall be determined in accordance with Section 2.5.2.

2.5.3.1 Calculating Cover: The concrete cover for an individual tendon is the minimum thickness of concrete between the surface of the tendon and the fire-exposed surface of the beam; except for ungrouted ducts, the assumed cover thickness is the minimum thickness of concrete between the surface of the duct and the surface of the beam. For beams in which several tendons are used, the cover, for purposes of Table 2-7 (a) and (b), is the average of the minimum cover of the individual tendons. For corner tendons, the minimum cover used in the calculation shall be one-half the actual value. For stemmed members with two or more prestressing tendons located along the vertical centerline of the member, the average cover shall be the distance from the bottom of the member to the centroid of the tendons. The actual cover for an individual tendon shall not be less than one-half the value shown in Table 2-7 (a) or (b) or 1 in. (25 mm), whichever is greater.

2.6 Reinforced Concrete Columns

The least dimension of reinforced concrete columns of different types of concrete for fire resistance ratings of 1 to 4 hours shall be not less than the dimensions indicated in Table 2-8.

2.6.1 Minimum Cover for Reinforcement

The minimum thickness of concrete cover to the main longitudinal reinforcement in columns, regardless of the type of aggregate used in the concrete, shall not be less than 1 in. (25 mm) times the number of hours of required fire resistance, or 2 in. (51 mm), whichever is less.

2.6.2 Columns Built into Walls

The minimum dimensions of Table 2-8 shall not apply to a reinforced concrete column that is built into a concrete or masonry wall provided all of the following are met:

1. the fire resistance of the wall is equal to or greater than the required rating of the column;
2. openings in the wall are protected in accordance with the general building code so that no more than one face of the column will be exposed to fire at the same time; and
3. the main longitudinal reinforcement in the column has cover of not less than required by Section 2.6.1.

3. STANDARD METHODS FOR DETERMINING THE FIRE RESISTANCE OF TIMBER AND WOOD STRUCTURAL ELEMENTS

3.1 Scope

Chapter 3 contains methodologies for determining the fire performance of large section timbers and wood structural elements. The procedure for determining the fire resistance of a heavy timber member is based on a mathematical model and on testing that demonstrated the intrinsic ability of larger wood members to sustain a structural load during severe fire exposure. The second part of this Chapter contains the Component Additive Method (CAM), which utilized tests and Harmathy's rules for fire endurance rating.

3.1.1 Limitations

These calculation methods are applicable to fire endurance times of up to 1 hour.

3.1.2 Dimensions and Metric Conversion

In this Chapter, the section dimensions of structural members are nominal when expressed in in.-lb units, unless it is explicitly stated that actual dimensions are specified. The corresponding metric units are

TABLE 2-8. Minimum Concrete Column Dimension

Concrete Aggregate Type	Minimum Column Dimension									
	1 hr		1.5 hr		2 hr		3 hr		4 hr	
	in.	mm	in.	mm	in.	mm	in.	mm	in.	mm
Siliceous	8	203	9	229	10	254	12	305	14	356
Carbonate	8	203	9	229	10	254	11	279	12	305
Sand-lightweight	8	203	8.5	216	9	229	10.5	267	12	305

given in parentheses and are always for the actual dimensions. Thicknesses of panel products and dimensions other than those of sections are actual values in both systems of units.

3.2 Notations and Definitions

γ = constant (2.54 min/in. or 0.1 min/mm)

b = actual breadth (width) of a beam or actual larger side of a column before exposure to fire, measured in inches or mm

d = actual depth of a beam or actual smaller side of a column before exposure to fire, measured in inches or mm

z = load factor; see Section 3.2.2.1

l = the unsupported length of a column, measured in inches or mm

r = ratio of applied load to allowable load, indicated as a percentage

K_e = effective length; see Section 3.2.2.1.

Large Timber Section: a structural wood element having minimum cross-sectional dimensions of 6 in. $\times$ 6 in. (nominal).

3.3 Design of Fire-Resistive Exposed Wood Members

3.3.1 Analytical Method for Exposed Wood Members

Fire resistance ratings of up to 1-hour's duration, for exposed timber members, shall be determined in accordance with the criteria contained in this Section and Section 3.3.2.2 for beams and Section 3.3.2.3 for columns.

Applied loads on beams and columns shall be determined in accordance with the provisions approved by the authority having jurisdiction. In the absence of loading provisions, loads shall be determined using ASCE 7-95, *Minimum Design Loads for Buildings and Other Structures*. Allowable loads on beams and columns shall be determined in accordance with the provisions of the applicable building code. In the absence of design provisions, allowable loads shall be determined from the NDS.[10]

3.3.2.1 Load Factor and Effective Length Factor:

The load factor, z, shall be determined from the equations given in Sections 3.3.2.1.1 and 3.3.2.1.2. The effective length factor shall be determined from Table G1 in NDS-91, the 1991 Edition of the *National Design Specification for Wood Construction*.

3.3.2.1.1 Load Factor of Short Columns: For columns having a $K_e l/d \leq 11$, z shall be equal to 1.5 when the

load ratio, r is equal to or less than 50%. When r is greater than 50%, z shall be equal to:

$$z = 0.9 + 30/r \qquad (3.1)$$

3.3.2.1.2 Load Factor of Beams and Other Columns: For all beams and other columns ($K_e l/d > 11$), z shall be equal to 1.3 when r is equal to or less than 50%. When r is greater than 50%, z shall be equal to:

$$z = 0.7 + 30/r \qquad (3.2)$$

3.3.2.2 Beams: The calculation for the fire resistance rating (or time to failure, t), measured in minutes, of timber beams with a least nominal dimension of 6 in. (140 mm) shall be equal to:

$$t = yzb\,[4-2(b/d)] \qquad (3.3)$$

for beams exposed to fire on four sides; or,

$$t = yzb\,[4-(b/d)] \qquad (3.4)$$

for beams exposed to fire on three sides.

3.3.2.2.1 Glued Laminated Timber Beams: The tension zone shall be increased by removing one core lamination, moving the tension zone inward, and adding the equivalent of an additional nominal 2-in.-thick (38 mm) outer tension lamination as illustrated in Figure 3-1.

3.3.2.3 Columns: The calculation for the fire resistance, measured in minutes, of timber columns with a minimum dimension of 6 in. (140 mm), shall be as calculated by Equation (3.5) or (3.6). Equation (3.6) for columns exposed to fire on three sides shall apply only when the unexposed face is the smaller column dimension.

$$t = \gamma zd\,[3-(d/b)] \qquad (3.5)$$

for columns exposed to fire on four sides; or

$$t = \gamma zd\,[3-(d/2b)] \qquad (3.6)$$

for columns exposed to fire on three sides

3.3.2.4 Connectors and Fasteners: Where a 1-hour fire resistance rating is required, connectors and fasteners shall be protected from fire exposure by 1.5 in. (38 mm) of wood, 5/8-in. (16 mm) Type X gypsum board, or other approved materials.

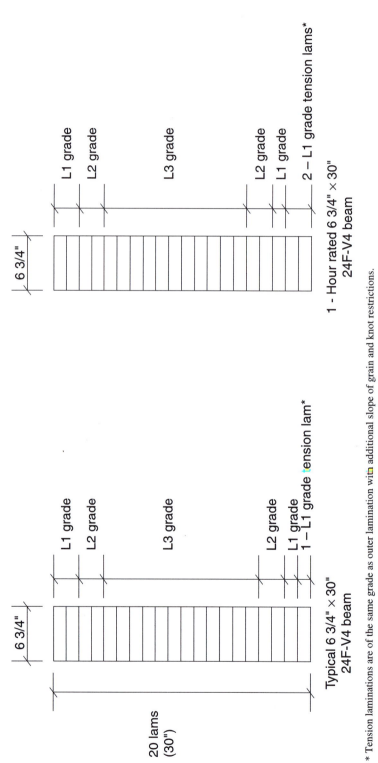

20 lams
(30")

L1 grade
L2 grade

L3 grade

L2 grade
L1 grade
1 – L1 grade tension lam*

Typical 6 3/4" × 30"
24F-V4 beam

6 3/4"

L1 grade
L2 grade

L3 grade

L2 grade
L1 grade
2 – L1 grade tension lams*

1 - Hour rated 6 3/4" × 30"
24F-V4 beam

6 3/4"

* Tension laminations are of the same grade as outer lamination with additional slope of grain and knot restrictions.

FIGURE 3-1 Effect of Relocating a Tension Laminate into the Core Zone

3.4 Component Additive Method for Calculating and Demonstrating Assembly Fire Endurance

3.4.1 Analytical Method for Protected Wood-Frame Assemblies

Fire resistance ratings up to 1 hour's duration are permitted to be calculated for walls, floor/ceiling, and roof/ceiling assemblies by combining the individual component times of the assembly in accordance with this section. The calculated time shall equal or exceed the required fire resistance rating of the assembly.

3.4.1.1 Component Times: The fire resistance rating of a wood-frame assembly shall equal the sum of the time assigned to the membrane on the fire exposed side, Table 3-1, the time assigned to the framing members, from Table 3-2, and the time assigned for additional contribution by cavity insulation (for walls and partitions), from Table 3-3. The membrane(s) on the fire-exposed side shall be attached to the framing members in accordance with the applicable building code requirements. The membrane on the unexposed side shall not be included in determining the fire resistance of the assembly. The time to be assigned to a masonry veneer of an exterior wall, if required to be rated for fire exposure from the exterior side, shall be

TABLE 3-1. Time Assigned to Protective Membranes

Description of Finish	Time (min)
3/8 in. (9.5 mm) Douglas fir plywood phenolic bonded	5
1/2 in. (12.7 mm) Douglas fir plywood phenolic bonded	10
5/8 in. (15.9 mm) Douglas fir plywood phenolic bonded	15
3/8 in. (9.5 mm) gypsum board	10
1/2 in. (12.7 mm) gypsum board	15
5/8 in. (15.9 mm) gypsum board	20
1/2 in. (12.7 mm) Type X gypsum board	25
5/8 in. (15.9 mm) Type X gypsum board	40
Double 3/8 in. (9.5 mm) gypsum board	25
1/2 + 3/8 in. (12.7 mm + 9.5 mm) gypsum board	35
Double 1/2 in. (12.7 mm) gypsum board	40
Double 5/8 in. (15.9 mm) Type X gypsum board	55

Note: On walls, gypsum board shall be installed with the long dimension (edge) parallel to framing members with all face layer joints and fasteners finished. 5/8 in. (15.9 mm) Type X gypsum board is permitted to be installed horizontally with the horizontal joints unsupported. On floor/ceiling or roof/ceiling assemblies, gypsum board shall be installed with the long dimension at right angles to framing members and shall have all face layer joints and fasteners finished.

TABLE 3-2. Time Assigned to Wood-Frame Components

Description of Frames	Time (min)
Wood studs minimum 2 in. nominal (38 mm), 16 in. (406 mm) on center	20
Wood joists minimum 2 in. nominal (38 mm), 16 in. (406 mm) on center	10
Wood roof and floor truss assemblies, 24 in. (610 mm) on center	5

TABLE 3-3. Time Assigned for Insulation of Cavity

Description of Additional Protection	Time (min)
Add to the fire endurance rating of wood stud walls if the spaces between the studs are filled with rockwool or slag mineral wool batts weighing not less than 1 lb/ft² of wall surface (4.8 kg/m²)	15
Add to the fire endurance rating of non-load bearing wood stud walls if the spaces between the studs are filled with glass fiber batts weighing not less than 0.6 lb/ft² of wall surface (2.9 kg/m²)	5

determined according to the methods described in Chapter 4.

3.4.1.2 Exposed Plywood: For a wall or partition where only plywood is used as the membrane on the side that would be exposed to the fire, insulation in accordance with Table 3-3 shall be used within the assembly. No credit shall be given for the insulation.

3.4.1.3 Unsymmetrical Wall Assemblies: When dissimilar membranes are used on opposite faces of a wall assembly, the fire endurance is permitted to be determined based on the calculations for the least fire-resistant side.

Exception: Where exterior walls are required to be rated for exposure to fire from the inside only, the non-fire side (exterior) membrane shall be constructed of any combination of materials listed in Table 3-4 or any other membrane listed at 15 minutes or greater in Table 3-1.

3.4.1.4 Floor/Ceiling and Roof/Ceiling Assemblies: Fire-resistant floor/ceiling and roof/ceiling assemblies shall have an upper membrane (unexposed side) provided in accordance with Table 3-5.

As an alternative to the unexposed (upper) membranes listed in Table 3-5, combinations of membranes

TABLE 3-4. Membrane on Exterior Face of Walls

Sheathing	Paper	Exterior Finish
5/8 in. (15.9 mm) T&G lumber 5/16 in. (7.9 mm) exterior grade plywood 1/2 in. (12.7 mm) gypsum sheathing	Sheathing paper per building code	Lumber siding Wood shingles and shakes 1/4 in. (6.4 mm) exterior grade plywood 1/4 in. (6.4 mm) hardboard Metal siding Stucco on metal lath Masonry veneer
None	None	3/8 in. (9.5 mm) exterior grade plywood

TABLE 3-5. Flooring or Roofing Membrane

Assembly	Structural Members	Subfloor or Roof Deck	Finish Flooring or Roofing
Floor	Wood	1/2 in. (12.7 mm) plywood or 11/16 in. (17.5 mm) T&G softwood lumber	Hardwood or softwood flooring on building paper; or resilient flooring, parquet floor, felted-synthetic-fiber floor coverings, carpeting, or ceramic tile on 3/8 in. (9.5 mm) thick panel-type underlay; or ceramic tile on 1 1/4 in. (31.8 mm) mortar bed
Roof	Wood	1/2 in. (12.7 mm) plywood or 11/16 in. (17.5 mm) T&G softwood lumber	Finish roofing material with or without insulation

from Table 3-1 with a time assigned value of at least 15 minutes shall be permitted.

4. STANDARD CALCULATION METHODS FOR DETERMINING THE FIRE RESISTANCE OF MASONRY

4.1 Scope

Calculated acceptable fire resistance ratings of masonry are to be determined in accordance with the provisions of this Chapter. Except where the provisions of this Chapter are more stringent, the design, construction, and material requirements of masonry including units, mortar, grout, control and expansion joint materials, and reinforcement shall comply with ACI 530–95, *Building Code Requirements for Masonry Structures* [4.2.1].

4.2 Definitions

Words and terms used in this Chapter shall have the following meanings:

Column, masonry: An isolated vertical member whose horizontal dimension measured at right angles to the thickness does not exceed three times its thickness and whose height is at least three times its thickness.

Control joint: A continuous vertical or transverse joint placed between units that is used in concrete masonry to create a plane of weakness that, used in conjunction with reinforcement or joint reinforcement, controls the location of cracks due to volume changes resulting from shrinkage or creep.

Equivalent thickness: The average thickness of solid material in the wall.

Expansion joint: A continuous vertical or transverse void placed between units that is used to separate clay brick masonry into segments to prevent cracking due to changes in temperature, moisture and freezing expansion, elastic deformation, and creep due to loads.

Grout: A mixture of cementitious material and aggregate to which sufficient water is added to produce pouring consistency without segregation of the constituents.

Lintel: A beam placed over an opening in a wall.

Loose fill material: A material used to completely fill the designated vertically aligned cells of hollow masonry units or the space between wythes in a cavity wall.

Masonry unit, clay: A unit made of clay or shale, usually formed into a rectangular prism while in the plastic state and burned or fired in a kiln.

Masonry unit, concrete: A building unit or block made of cement and suitable aggregates.

Masonry unit, hollow: A masonry unit whose net cross-sectional area in any plane parallel to the bearing surface is less than 75% of its gross cross-sectional area measured in the same plane.

Masonry unit, solid: A masonry unit whose net cross-sectional area in every plane parallel to the bearing surface is 75% or more of its gross cross-sectional area measured in the same plane.

Mortar: A plastic mixture of cementitious materials, fine aggregates, and water used to bond masonry or other structural units.

Wall: A vertical element with a horizontal length-to-thickness ratio greater than three.

Wall, composite: A wall in which at least one of the wythes is dissimilar to the other wythe or wythes with respect to the type or grade of masonry unit or mortar; a wall with at least two dissimilar wythes separated by a collar joint and composed of clay or concrete masonry units, concrete or mixture thereof.

4.3 Equivalent Thickness

The equivalent thickness of masonry shall be determined in accordance with the provisions of this Section.

4.3.1 Hollow Unit Masonry

The equivalent thickness of masonry, T_{ea}, shall be based on the equivalent thickness of the masonry unit, T_e, determined in accordance with ASTM C140 for concrete masonry units or ASTM C67 for clay masonry units, plus the equivalent thickness of finishes, T_{ef} not considered in Sections 4.3.2 and 4.3.3 (see Section 4.4.1).

$$T_{ea} = T_e + T_{ef} \qquad (4.1)$$

$$T_e = V_n/LH \qquad (4.2)$$

where

T_e = equivalent thickness of masonry unit, in. (mm)
T_{ea} = equivalent thickness of masonry, in. (mm)
T_{ef} = equivalent thickness of finishes, in. (mm)
L = specified length of masonry unit, in. (mm)
V_n = net volume of masonry unit, in.^{3} (mm^3)
H = specified height of masonry unit, in. (mm)

4.3.2 Solid Grouted Construction

The equivalent thickness, T_e, of solidly grouted mansory units shall be the specified thickness of the unit.

4.3.3 Air Spaces or Cells Filled with Loose Fill Material

The equivalent thickness of filled air spaces or hollow masonry shall be taken as the specified thickness of the unit where loose fill materials are sand, pea gravel, crushed stone, or slag conforming to ASTM C33; pumice, scoria, expanded shale, expanded clay, expanded slate, expanded slag, expanded fly ash, or cinders conforming to ASTM C331; or perlite or vermiculite conforming to ASTM C332.

4.4 Masonry Walls

The minimum equivalent thickness of various types of plain or reinforced masonry bearing or non-bearing walls required to provide fire resistance ratings of 1/2 hour to 4 hours is indicated in Table 4-1.

Table 4-1 is applicable only to those aggregates for which the fire-resistance properties have been tested and documented.

4.4.1 Walls with Gypsum Wallboard or Plaster Finishes

The fire resistance rating of masonry walls with gypsum wallboard or plaster finishes applied to one or two sides of the wall shall be determined using the provisions of this Section.

4.4.1.1 Calculation for Non–Fire-Exposed Side: Where the finish of gypsum wallboard or plaster is applied to the non–fire-exposed side of the wall, the fire resistance rating of the entire assembly shall be determined as follows: The thickness of the finish shall be adjusted by multiplying the actual thickness of the finish by the applicable factor from Table 4-2 based on the type of clay masonry unit or concrete masonry unit. The adjusted finish thickness shall be added to the equivalent thickness of masonry wall and the fire resistance rating of the masonry wall, including the finish, determined from Table 4-1. Where a finish is added to the non–fire-exposed side of a hollow clay masonry unit, no increase in fire resistance shall be permitted.

4.4.1.2 Calculation for Fire-Exposed Side: Where the finish of gypsum wallboard or plaster is applied to the fire-exposed side of the wall, the fire resistance rating of the entire wall shall be determined as follows: The time assigned to the finish by Table 4-3 shall be added to the fire resistance rating determined from Table 4-1 for the masonry wall alone or to the rating determined in accordance with Section 4.4.1.1 for the masonry wall and finish on the non–fire-exposed side.

TABLE 4-1. Fire Resistance Rating of Masonry

Minimum Required Equivalent Thickness of the Masonry, [1,3] in.

Aggregate Type in the Concrete Masonry Unit[2]	Fire Resistance Rating Period													
	4 hr		3 hr		2 hr		1.5 hr		1 hr		3/4 hr		1/2 hr	
	in.	mm	in.	mm	in.	mm	in.	mm	in.	mm	in.	mm	in.	mm
Calcareous or siliceous gravel	6.2	157	5.3	135	4.2	107	3.6	91	2.8	71	2.4	61	2.0	51
Limestone, cinders, or slag	5.9	150	5.0	127	4.0	102	3.4	86	2.7	69	2.3	58	1.9	48
Expanded clay, shale, or slate	5.1	130	4.4	112	3.6	91	3.3	84	2.6	66	2.2	56	1.8	46
Expanded slag or pumice	4.7	119	4.0	102	3.2	81	2.7	69	2.1	53	1.9	48	1.5	38
Clay Masonry Unit														
Brick of clay or shale, unfilled	5.0	127	4.3	109	3.4	86	2.85	72	2.3	58	2.0	51	1.7	43
Brick of clay or shale, grouted or filled with perlite, vermiculite, or expanded shale aggregate	6.6	168	5.5	140	4.4	112	3.7	94	3.0	76	2.65	67	2.3	58

[1]Fire resistance rating between the hour fire resistance ratings listed shall be determined by linear interpolation based on the equivalent thickness value of the masonry.
[2]Minimum required equivalent thickness corresponding to the hour fire resistance rating for units made with a combination of aggregates shall be determined by linear interpolation based on the percent by volume of each aggregate used in the manufacture.
[3]Where combustible members are framed in the wall, the thickness of solid material between the end of each member and the opposite face of the wall, or between members set in from opposite sides, shall not be less than 95% of the thickness shown in the table.

TABLE 4-2. Multiplying Factors for Finishes on Non–Fire-Exposed Side of Masonry Walls

Finish Material	Type of Masonry			
	Solid Clay Brick	Clay Tile; Concrete Masonry Units of Expanded Shale and <20% Sand	Concrete Masonry Units of Expanded Shale or of Pumice, Expanded Slag, Expanded Clay, and <20% Sand	Concrete Masonry Units of Expanded Slag, Expanded Clay, or Pumice
Portland cement-sand plaster, lime sand plaster	1	0.75	0.75	0.5
Gypsum-sand plaster	1.25	1	1	1
Gypsum wallboard	3	2.25	2.25	2.25
Vermiculite or perlite aggregate plaster	1.75	1.5	1.25	1.25

4.4.1.3 Assume Each Side of Wall Is Fire-Exposed:
Where a wall is required to be fire resistance rated from both sides and has no finish on one side, or has different types or thicknesses of finishes on each side, the calculation procedures of Sections 4.4.1.1 and 4.4.1.2 shall be performed twice (i.e., with each side of the wall as the fire-exposed side). The fire resistance rating of the wall, including finishes, shall not exceed the lower of the two values calculated.

4.4.1.4 Minimum Rating Provided by Masonry Walls:
Where the finish applied to a masonry wall contributes to the fire resistance rating, the masonry wall alone shall provide not less than one-half of the total required fire resistance rating. The contribution to the fire resis-

TABLE 4-3. Time Assigned to Finish Materials on Fire-Exposed Side of Masonry Wall

Finish Description	Time (min)
Gypsum Wallboard	
3/8 in. (9.5 mm)	10
1/2 in. (12.7 mm)	15
5/8 in. (15.9 mm)	
2 layers of 3/8 in. (9.5 mm)	25
1 layer of 3/8 in. (9.5 mm), 1 layer of 1/2 in. (12.7 mm)	35
2 layers of 1/2 in. (12.7 mm)	40
Type X Gypsum Wallboard	
1/2 in. (12.7 mm)	25
5/8 in. (15.9 mm)	40
Portland Cement-Sand Plaster Applied Directly to Masonry	*
Portland Cement-Sand Plaster on Metal Lath	
3/4 in. (19.0 mm)	20
7/8 in. (22.2 mm)	25
1 in. (25.4 mm)	30
Gypsum Sand Plaster on 3/8 in. (9.5 mm) Gypsum Lath	
1/2 in. (12.7 mm)	35
5/8 in. (15.9 mm)	40
3/4 in. (19.0 mm)	50
Gypsum Sand Plaster on Metal Lath	
3/4 in. (19.0 mm)	50
7/8 in. (22.2 mm)	60
1 in. (25.4 mm)	80

* The actual thickness of portland cement-sand plaster shall only be included in the determination of the equivalent thickness of masonry for use in Table 4-1 when it is 5/8 in. (15.9 mm) or less in thickness.

tance of the finish on the non–fire-exposed side of the wall shall not exceed 0.5 times the contribution of the masonry alone.

4.4.1.5 Installation of Finishes: Finishes on masonry walls that contribute to the total fire resistance rating of the wall shall comply with the installation requirements of this section and applicable provisions of the legally adopted building code.

4.4.1.5.1 Gypsum Wallboard or Gypsum Lath and Plaster: For adding fire resistance to masonry wall assemblies, gypsum wallboard and gypsum lath shall be attached in accordance with one of the three methods in Sections 4.4.1.5.1(a) through 4.4.1.5.1(c).

4.4.1.5.1(a) Self-tapping drywall screws spaced at a maximum of 12 in. (305 mm) on center and shall penetrate 3/8 in. (10 mm) into resilient steel furring channels running horizontally and spaced at a maximum 24 in. (610 mm) on center.

4.4.1.5.1(b) Lath nails shall be spaced at a maximum of 12 in. (305 mm) on center and shall penetrate 3/4 in. (19 mm) into nominal 1 in. $\times$ 2 in. (25 mm $\times$ 50 mm) wood furring strips that are secured to the masonry by 2 in. (50 mm) concrete nails spaced a maximum of 16 in. (406 mm) on center. The center to center spacing of wood furring strips shall not exceed 16 in. (406 mm).

4.4.1.5.1(c) A 3/8 in. (10 mm) bead of panel adhesive shall be placed around the perimeter of the wallboard and across the diagonals. After the wallboard is laminated to the masonry surface, the wallboard shall be secured with one masonry nail for each 2 ft^2 (0.18 m^2) of panel.

4.4.1.5.2 Plaster and Stucco: Application of plaster and stucco finishes to increase the fire resistance rating to the surface of masonry shall be in accordance with the provisions of the legally adopted building code.

4.4.2 Single-Wythe Wall Assemblies
The fire resistance rating of single-wythe masonry walls shall be in accordance with Table 4-1.

4.4.3 Multi-Wythe Wall Assemblies
The fire resistance rating of multi-wythe walls (Figure 4-1) of the same or dissimilar materials, with or without an air space between wythes, shall be based on the following equation:

$$R = (R_1^{0.59} + R_2^{0.59} + ... + R_n^{0.59} + A_1 + A_2 + ... A_n)^{1.7} \qquad (4.3)$$

where

$R_1, R_2, ... R_n$ = fire resistance rating of wythe 1, 2, ... n, respectively (hr)

$A_1, A_2, ... A_n$ = 0.30; factor for one air space, having a width of 1/2 in. (12.7 mm) to 3 1/2 in. (88.9 mm) between wythes; 0.60; factor for two air spaces, each having a width of 1/2 in. (12.7 mm) to 3 1/2 in. (88.9 mm) between wythes

4.4.4 Multi-Wythe Walls with Dissimilar Materials
For multi-wythe walls consisting of two or more wythes of dissimilar materials (concrete, concrete masonry units, or clay masonry units), the determination of the fire resistance periods of the dissimilar wythes, R_n, shall be in accordance with Section 3.1 for concrete and Section 4.4 for concrete masonry units or clay masonry units.

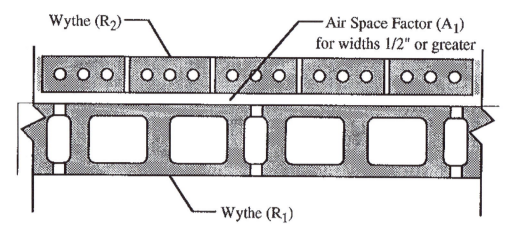

*Note: Partially filled collar joints are not considered air spaces

R_1 = Fire resistance rating of wythe 1
R_2 = Fire resistance rating of wythe 2
A_1 = Air space factor = 3.3 for one air space; 6.7 for two air spaces

FIGURE 4-1. Fire Resistance of Multi-Wythe Masonry Wall

TABLE 4-4. Reinforced Masonry Columns

Minimum Column Dimensions							
1 hr		2 hr		3 hr		4 hr	
in.	mm	in.	mm	in.	mm	in.	mm
8	203	10	254	12	305	14	356

TABLE 4-5. Reinforced Masonry Lintels Minimum Longitudinal Reinforcing Cover

	Fire Resistance Rating							
	1 hr		2 hr		3 hr		4 hr	
Nominal Lintel Width	in.	mm	in.	mm	in.	mm	in.	mm
6 in. (152 mm)	1 1/2	38	2	51	—	—	—	—
8 in. (203 mm)	1 1/2	38	1 1/2	38	1 3/4	44	3	76
10 in. (254 mm)	1 1/2	38	1 1/2	38	1 1/2	38	1 3/4	44

4.4.5 Movement Joints

Control joints for concrete masonry and expansion joints for clay masonry in fire-rated wall assemblies shall be in accordance with Figure 4-2. Where walls are permitted to have unprotected openings, fire-resistant joints are not required.

4.5 Reinforced Masonry Columns

The fire resistance rating of reinforced masonry columns shall be based on the least plan dimension of the column in accordance with the requirements of Table 4-4. The minimum cover of longitudinal reinforcement shall be 2 in. (50 mm).

4.6 Masonry Lintels

The fire resistance rating of masonry lintels shall be determined based on the nominal thickness of the lintel and the minimum cover of longitudinal reinforcement in accordance with Table 4-5.

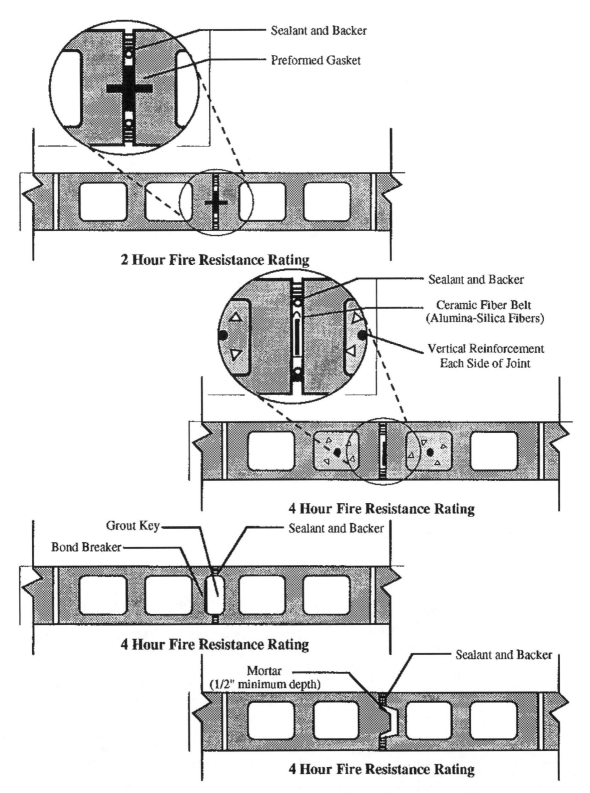

FIGURE 4-2. Movement Joints for Fire-Resistant Masonry Assemblies

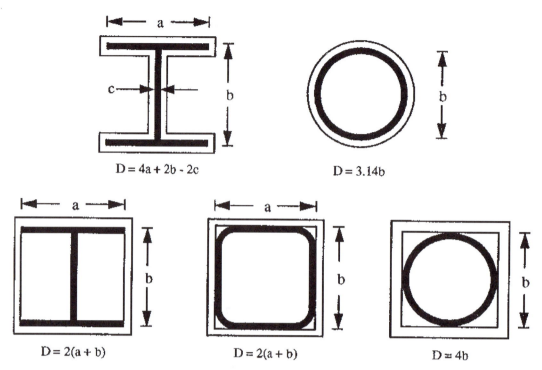

FIGURE 5-1. Determination of the Heated Perimeter (D) of Steel Columns

5. STANDARD METHODS FOR DETERMINING THE FIRE RESISTANCE OF STRUCTURAL STEEL CONSTRUCTION

5.1 Scope

This Section describes analytical procedures for determining the fire resistance rating of steel columns, beams, girders, and trusses protected as specified.

5.2 Structural Steel Columns

This Section describes procedures for determining the fire resistance rating of column assemblies. In general, these procedures are based on the weight (*W*) and heated perimeter (*D*) of steel columns. As used in these sections, *W* is the average weight of a structural steel column in pounds per linear foot (kilograms per meter). The heated perimeter (*D*) is the inside perimeter of the fire-protection material in inches (millimeters) as illustrated in Figure 5-1. Other terms are specifically defined in the applicable section(s).

5.2.1 Gypsum Wallboard

The fire resistance rating of wide-flange, pipe and tubular structural steel columns with weight-to-heated-perimeter ratios (*W/D*) less than or equal to 3.65 (customary units) or 0.215 (SI units) protected with Type X gypsum wallboard shall be determined in accordance with either of the following expressions:

$$R = 2.17 \left[\frac{h(W'/D)}{2} \right]^{0.75} \tag{5.1}$$

In SI units

$$R = 1.60 \left[\frac{h(W'/D)}{2} \right]^{0.75} \tag{5.2}$$

where

R = fire resistance in hours

h = total thickness of gypsum wallboard in inches (millimeters)

D = heated perimeter of the structural steel column in inches (millimeters)

W' = total weight of the structural steel column and gypsum wallboard protection in pounds per linear foot (kilograms per meter).

$$W' = W + \frac{50hD}{144}$$

In SI units

$$W' = W + 0.0008hD$$

The gypsum wallboard shall be supported and fastened as illustrated in either Figure 5-2, for fire resis-

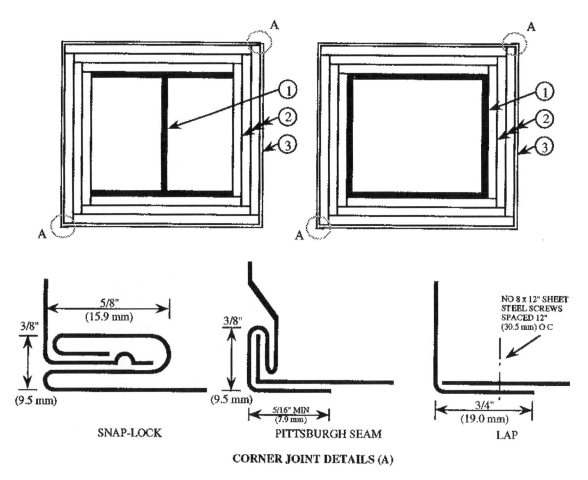

CORNER JOINT DETAILS (A)

Notes:
1. Structural steel column, either wide-flange or tubular shapes.
2. Type X gypsum wallboard. For single-layer applications, the wallboard shall be applied vertically with no horizontal joints. For multiple-layer applications, horizontal joints are permitted at a minimum spacing of 8 feet (2.4 m), provided that the joints in successive layers are staggered at least 12 inches (304.8 mm). The total required thickness of wallboard shall be determined on the basis of the specified fire-resistance rating and the weight and heated perimeter of the column. For fire resistance ratings of two hours or less, one of the required layers of gypsum wallboard may be applied to the exterior of the sheet steel column covers with 1-inch (25.4-mm) long Type S screws spaced 1 inch (25.4 mm) from the wallboard edge and 8 inches on center. For such installations, 0.016 inch (0.4 mm) minimum-thickness galvanized steel corner beads with 1 1/2 inch (203.2 mm) legs shall be attached to the wallboard with Type S screws spaced 12 inches (304.8 mm) on center.
3. For fire resistance ratings of three hours or less, the column covers shall be fabricated from 0.024 inch (0.6 mm) minimum-thickness galvanized or stainless steel. For four-hour fire resistance ratings, the column covers shall be fabricated from 0.024 inch (0.6 mm) minimum-thickness stainless steel. The column covers shall be erected with the snap lock or pittsburgh joint details. For fire resistance ratings of two hours or less, column covers fabricated from 0.027 inch (0.7 mm) minimum-thickness galvanized or stainless steel may be erected with lap joints. The lap joints may be located anywhere around the perimeter of the column cover. The lap joint shall be secured with 1/2-inch (12.7-mm) long No. 8 sheet metal screws spaced 12 inches (304.8 mm) on center. The column covers shall be provided with a minimum expansion clearance of 1/8 inch per linear foot (10.4 mm/m) between the ends of the cover and any restraining construction.

FIGURE 5-2. Gypsum Wallboard Protected Structural Steel Columns with Sheet Steel Column Covers (Four Hours or Less)

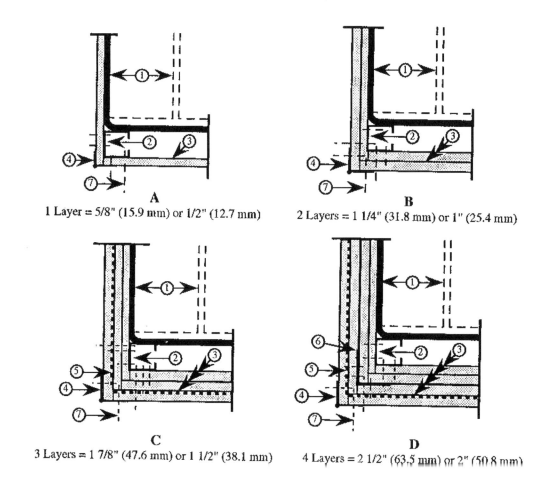

A

1 Layer = 5/8" (15.9 mm) or 1/2" (12.7 mm)

B

2 Layers = 1 1/4" (31.8 mm) or 1" (25.4 mm)

C

3 Layers = 1 7/8" (47.6 mm) or 1 1/2" (38.1 mm)

D

4 Layers = 2 1/2" (63.5 mm) or 2" (50.8 mm)

Notes:

1. Structural steel column, either wide-flange, pipe, or tubular shapes.

2. 1 5/8-inch (15.9-mm) deep studs fabricated from 0.021 inch (0.5 mm) minimum-thickness galvanized steel with 1 5/16 (33.3 mm) or 1 7/16 inch (36.5 mm) legs and 1/4 inch (6.4 mm) stiffening flanges. The length of the steels studs shall be 1/2 inch (12.7 mm) less than the height of the assembly.

3. Type X gypsum wallboard. For single-layer applications, the wallboard shall be applied vertically with no horizontal joints. For multiple-layer applications, horizontal joints shall be permitted at a minimum spacing of 8 feet (2.4 m), provided that the joints in successive layers are staggered at least 12 inches (304.8 mm). The total required thickness of wallboard shall be determined on the basis of the specified fire-resistance rating and the weight and heated perimeter of the column.

4. Galvanized steel corner beads (0.016 inch [0.4 mm] minimum thickness) with 1 1/2 inch (38.1 mm) legs attached to the wallboard with 1-inch (25.4-mm) long Type S screws spaced 12 inches (304.8 mm) on center.

5. No. 18 SWG steel tie wire spaced 24 inches (160 mm) on center.

6. Sheet metal angles with 2 inch (50.8 mm) legs fabricated from 0.021 inch (0.5 mm) minimum-thickness galvanized steel.

7. Type S screws 1-inch (25.4-mm) long shall be used for attaching the first layer of wallboard to the steel studs and the third layer to the sheet metal angles at 24 inches (609.6 mm) on center. Type S screws 1 3/4- inches (44.5-mm) long shall be used for attaching the second layer of wallboard to the steel studs and the fourth layer to the sheet metal angles at 12 inches (304.8 mm) on center. Type S screws 2-1/4 inches (57.2-mm) long shall be used for attaching the third layer of wallboard to the steels studs at 12 inches (304.8 mm) on center.

FIGURE 5-3. Gypsum Wallboard Protected Structural Steel Columns with Steel Stud/Screw Attachment System (Three Hours or Less)

tance ratings of 4 hours, or less, or Figure 5-3, for fire resistance ratings of 3 hours or less. For structural steel columns with weight-to-heated-perimeter ratios (W/D) greater than 3.65 (customary units) or 0.215 (SI units), the thickness of gypsum wallboard required for specified fire resistance ratings shall be the same as the thickness determined for a W14 × 233 (customary units) or W360×347 (SI units) wide-flange shape.

5.2.2 Spray-Applied Materials

The fire resistance rating of structural steel columns protected with spray-applied fire protection materials, as illustrated in Figure 5-4, shall be determined in accordance with the following expressions:

$$R = \left(C_1 \frac{W}{D} + C_2 \right) h \quad \text{(wide-flange columns)} \quad (5.3)$$

In SI units

$$R = \left(C_3 \frac{W}{D} + C_4 \right) h$$

$$R = C_1' \left(\frac{A}{P} \right) h + C_2' \quad \text{(pipe and tubular columns)} \quad (5.4)$$

In SI units

$$R = C_3' \left(\frac{A}{P} \right) h + C_4' \quad \text{(pipe and tubular columns)}$$

where

R = fire resistance in hours

h = thickness of spray-applied fire protection material in inches (millimeters)

D = heated perimeter of the structural steel column in inches (millimeters)

W = average weight of the steel column in pounds per linear foot (kilograms per meter)

C_1, C_2, C_3, and C_4 = material-dependent constants for wide-flange columns

P = heated perimeter of the structural steel column in inches (millimeters)

A = cross-sectional area of the steel column in square inches (millimeters)

C_1', C_2', C_3', and C_4' = material-dependent constants for pipe and tubular columns

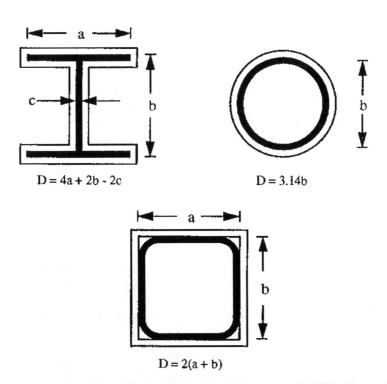

FIGURE 5-4. Structural Steel Columns with Spray-Applied Fire Protection

The material-dependent constants shall be determined for specific fire-protection materials on the basis of standard fire endurance tests conducted in accordance with ASTM E119.

Unless approved by the authority having jurisdiction, based on evidence substantiating a broader application, the use of these equations shall be limited by all of the following conditions:

1. Columns with weight-to-heated-perimeter ratios (*W/D*) that are equal to or greater than the smallest tested column.
2. Columns with weight-to-heated-perimeter ratios (*W/D*) that are equal to or less than the largest tested column.
3. Thicknesses of protection that are equal to or greater than the minimum tested thickness.
4. Thicknesses of protection that are equal to or less than the maximum tested thickness.
5. Ratings that are equal to or greater than the minimum fire resistance time for the applicable test series.
6. Ratings that are equal to or less than the maximum fire resistance time for the applicable test series.
7. The use of the wide-flange equation for other column geometries with open cross sections (i.e., channels, angles, and structural tees) shall be permitted. The use of the wide-flange equation shall not be permitted for columns with closed cross sections (i.e., pipe and tubular columns).
8. The use of the pipe and tubular equation shall be permitted for other column geometries with either open or closed cross sections.

5.2.3 Concrete-Filled Hollow Steel Columns

The fire resistance rating of hollow steel columns (e.g., pipe and tubular shapes) filled with unreinforced normal weight concrete shall be determined in accordance with the following expressions:

$$R = 0.58a \, \frac{(f'_c + 2.90)}{(KL - 3.28)} \, D^2 (D/C)^{0.5} \qquad (5.5)$$

In SI units

$$R = a \, \frac{(f'_c + 20)}{60 \, (KL - 1,000)} \, D^2 (D/C)^{0.5} \qquad (5.6)$$

where

R = fire resistance rating in hours
a = 0.07 for circular columns filled with siliceous aggregate concrete

= 0.08 for circular columns filled with carbonate aggregate concrete
= 0.06 for square or rectangular columns filled with siliceous aggregate concrete
= 0.07 for square or rectangular columns filled with carbonate aggregate concrete
f'_c = specified 28-day compressive strength of concrete in kips per square inch (megapascals)
KL = column effective length in feet (millimeters)
D = outside diameter for circular columns in inches (millimeters)
= outside dimension for square columns in inches (millimeters)
= least outside dimension for rectangular columns in inches (millimeters)
C = compressive force due to unfactored dead load and live load in kips (kilonewtons)

The application of these equations shall be limited by all of the following conditions:

1. The required fire resistance rating shall be less than or equal to 2 hours.
2. The specified compressive strength of concrete, f'_c, shall be equal to or greater than 2.90 kips per sq in. (20 megapascals).
3. The specified compressive strength of concrete, f'_c, shall not exceed 5.80 kips per sq in. (40 megapascals).
4. The column effective length shall be at least 6.50 ft (2,000 millimeters) and shall not exceed 13.0 ft (4,000 millimeters).
5. D shall be at least 5.50 in. (140 mm) and shall not exceed 12 in. (305 mm) for square and rectangular columns or 16 in. (410 mm) for circular columns.
6. C shall not exceed the design strength of the concrete core determined in accordance with AISC LRFD-94, *Load and Resistance Factor Design Specification for Structural Steel Buildings*.

5.2.4 Concrete or Masonry Protection

The fire resistance rating of non-composite structural steel columns protected with concrete, as illustrated in Figure 5-5, or masonry, as illustrated in Figure 5-6, shall be determined in accordance with the following expressions:

$$R = R_o \, (1 + 0.03 \, m) \qquad (5.7)$$

$$R_o = 0.17 \left(\frac{W}{D} \right)^{0.7} + 0.28 \, \frac{h^{1.6}}{k_c^{0.2}}$$
$$\times \left[1 + 26 \left(\frac{H}{\rho_c c_c h (L + h)} \right)^{0.8} \right] \qquad (5.8)$$

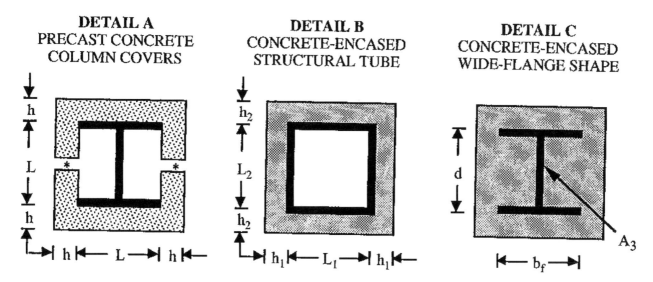

FIGURE 5-5. Concrete Protected Structural Steel Columns

Note: When the inside perimeter of the concrete protection is not square, L shall be taken as the average of L_1 and L_2. When the thickness of the concrete cover is not constant, h shall be taken as the average of h_1 and h_2.

* Joints shall be protected with a minimum 1 inch (25.4 mm) thickness of ceramic alumina-silica fiber insulation with a density of 4 to 8 pounds per cubic foot (64 to 128 kg/cu m). The thickness of the insulation shall not be less than one half the thickness of the column cover. The joint shall not exceed 1 inch (25.4 mm) maximum.

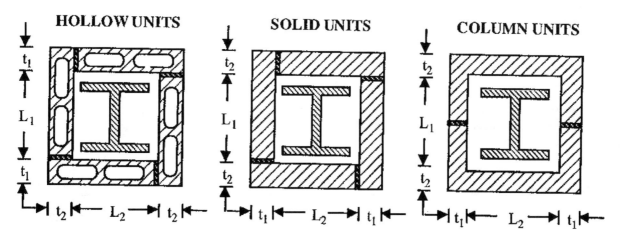

Note: The dimension L in equation 1.8 or 1.9 shall be the average of L_1 and L_2. The dimension h in Equation 1.8 or 1.9 shall be based on the equivalent thickness of the concrete masonry unit. For solid masonry units, h equals the lesser of t_1 and t_2. For hollow masonry units, h equals the lesser of t_1 and t_2, times the percent solid of the unit expressed as a decimal.

FIGURE 5-6. Concrete Masonry Protected Structural Steel Columns

In SI units

$$R_o = 1.22 \left(\frac{W}{D}\right)^{0.7} + 0.0027 \frac{h^{1.6}}{k_c^{0.2}}$$
$$\times \left[1 + 31,000 \left(\frac{H}{\rho_c c_c h(L + h)}\right)^{0.8}\right] \quad (5.9)$$

where

R = fire resistance rating at equilibrium moisture conditions in hours

R_o = fire resistance rating at zero moisture content in hours

TABLE 5-1. Properties of Concrete

	Normal Weight[1]	Structural Lightweight[2]
Thermal Conductivity, k_c	0.95 Btu/h-ft-°F (1.64 W/m-K)	0.35 Btu/h-ft-°F (0.61 W/m-K)
Specific Heat, c_c	0.20 Btu/h-ft-°F (0.84 kJ/kg-K)	0.20 Btu/h-ft-°F (0.84 kJ/kg-K)
Density, ρ_c	145 lb/ft³ (2,323 kg/m³)	110 lb/ft³ (1,762 kg/m³)
Moisture Content, m (percent by volume)	4	5

[1]Normal weight concrete is carbonate or siliceous aggregate concrete, as defined in Chapter 2.
[2]Structural lightweight concrete is lightweight or sand-lightweight concrete as defined in Chapter 2, with a minimum density (unit weight) of 110 lbs per cu ft (1,762 kg per cu m).

m = equilibrium moisture content of concrete or masonry, by volume (percent)

W = average weight of the steel column in pounds per linear foot (kilograms per meter)

D = heated perimeter of the steel column in inches (millimeters)

h = thickness of the concrete or equivalent thickness, for masonry in inches (millimeters)

k_c = ambient temperature thermal conductivity of concrete or masonry in Btu/hr-ft-°F (W/m-K)

H = ambient temperature thermal capacity of the steel column

= 0.11W Btu/ft-°F (kJ/m-K)

ρ_c = concrete or masonry density in pounds per cubic foot (kilograms per cubic meter)

c_c = ambient temperature specific heat of concrete or masonry in Btu/lb-°F (kJ/kg-K)

L = interior dimension of one side of square concrete or masonry box protection in inches (millimeters)

For wide-flange steel columns completely encased in concrete with all reentrant spaces filled, as shown in Figure 5-5, Detail C, the thermal capacity of the concrete within the reentrant spaces shall be added to the thermal capacity of the steel column in accordance with the following expressions:

$$H = 0.11W + \frac{\rho_c c_c}{144}(b_f d - A_s) \qquad (5.10)$$

In SI units

$$H = 0.46W + \frac{\rho_c c_c}{1,000,000}(b_f d - A_s) \qquad (5.11)$$

where

b_f = flange width of the steel column in inches (millimeters)

d = depth of the steel column in inches (millimeters)

A_s = cross-sectional area of the steel column in square inches (square millimeters)

TABLE 5-2. Properties of Concrete Masonry

Density		Thermal Conductivity	
(lb/ft³)	(kg/m³)	(Btu/h-ft-°F)	(W/m-K)
80	1,281	0.21	0.36
85	1,362	0.23	0.4
90	1,442	0.25	0.43
95	1,522	0.28	0.48
100	1,602	0.31	0.54
105	1,682	0.34	0.59
110	1,762	0.38	0.66
115	1,842	0.42	0.73
120	1,922	0.46	0.8
125	2,002	0.51	0.88
130	2,082	0.56	0.97
135	2,162	0.62	1.07
140	2,243	0.69	1.19
145	2,323	0.76	1.32
150	2,403	0.84	1.45

The specific heat, c_c, for concrete masonry shall be taken as 0.20 Btu/lb-°F (1.05 kJ/kg-K) and the equilibrium moisture content, m, as zero.

TABLE 5-3. Properties of Clay Masonry

Density		Thermal Conductivity	
(lb/ft³)	(kl/m³)	(Btu/h-ft-°F)	(W/m-K)
120	1,922	1.25	2.16
130	2,082	2.25	3.89

The specific heat, c_c, for clay masonry shall be taken as 0.24 Btu/16-°F (1.00 kJ/kg-K) and the equilibrium moisture content, m, as zero.

If specific data on the properties of concrete or masonry are not available, the values given in Tables 5-1, 5-2, and 5-3 shall be used.

5.3 Structural Steel Beams and Girders

This Section describes procedures for determining the fire resistance of structural steel beams and girders that differ in size from that specified in approved fire-resistant assemblies as a function of the thickness of fire protection material and the weight (W) and heated perimeter (D) of the beam or girder. As used in this Section, (W) is the average weight of the structural steel member in pounds per linear foot (kilograms per meter). The heated perimeter (D) is the inside perimeter of the fire protection material in inches (millimeters) as illustrated in Figure 5-7.

5.3.1 Spray-Applied Materials

The provisions in this Section shall apply to structural steel beams and girders protected with spray-applied cementitious or mineral fiber materials. Larger or smaller beams and girders shall be permitted to be substituted for the beams and girders specified in approved fire-resistant assemblies, provided that the thickness of fire protection material is adjusted in accordance with the following expressions:

$$h_2 = \left[\frac{W_1/D_1 + 0.60}{W_2/D_2 + 0.60} \right] h_1 \qquad (5.12)$$

In SI units

$$h_2 = \left[\frac{W_1/D_1 + 0.036}{W_2/D_2 + 0.036} \right] h_1 \qquad (5.13)$$

where

h = thickness of spray-applied fire protection materials in inches (millimeters)

W = weight of the structural steel beam or girder in pounds per linear foot (kilograms per meter)

D = heated perimeter of the structural steel beam or girder in inches (millimeters)

Note: Subscript 1 refers to the beam and fire protection thickness in the approved assembly; subscript 2 refers to the substitute beam or girder and the required thickness of fire protection material.

This equation shall be limited to beams with a weight-to-heated-perimeter ratio (W/D) of 0.37 or greater (customary units) or 0.022 (SI units). The thickness of protection shall not be less than 3/8 in. (9.5 millimeters).

5.4 Structural Steel Trusses

The fire resistance of structural steel trusses protected with cementitious or mineral fiber materials spray-applied to each of the individual members of a truss shall be determined in accordance with this Section. The thickness of protection for each truss member shall be determined in accordance with Section 5.2.2. The weight-to-heated-perimeter ratio (W/D) of truss members exposed to fire on all sides shall be determined on the same basis as columns, as specified in Section 5.2. The weight-to-heated-perimeter ratio (W/D) of truss members that directly support floor or roof construction shall be determined on the same basis as beams and girders, as specified in Section 5.3.

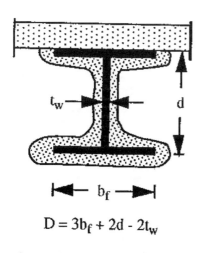

$$D = 3b_f + 2d - 2t_w$$

A. CONTOUR PROTECTION

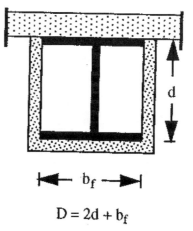

$$D = 2d + b_f$$

B. BOX PROTECTION

FIGURE 5-7. Determination of the Heated Perimeter of Steel Beams and Girders

COMMENTARY

This Commentary is not part of SEI/ASCE/SFPE 29-99 *Standard Calculation Methods for Structural Fire Protection*. It is included for information purposes.

This Commentary consists of explanatory and supplementary material designed to assist users of the Standard in applying the recommended requirements. In some cases, it will be necessary to adjust specific values in the Standard to local conditions; in others, a considerable amount of detailed information is needed to put the provisions into effect. This Commentary provides a place for supplying material that can be used in these situations and is intended to create a better understanding of the recommended requirements through brief explanations of the reasoning employed in arriving at them.

The sections of the Commentary are numbered to correspond to the sections of the Standard to which they refer. Since it is not necessary to have supplementary material for every section in the Standard, there are gaps in the numbering in the Commentary.

Commentary for 2: Standard Methods for Determining the Fire Resistance of Plain and Reinforced Concrete Construction

C2.1 Scope

Since around 1920, building codes in the United States have required that the fire resistance rating of construction assemblies be determined in accordance with ASTM E119, *Standard Test Methods for Fire Tests of Building Construction and Materials.*[1] Since then, literally thousands of small and full-scale fire tests have been performed on concrete assemblies. In addition, testing to determine the physical properties (e.g., compressive strength, modulus of elasticity, and yield strength) of concrete and reinforcing steel at elevated temperatures have been performed.

Much of this data was gathered early on, and fire testing over recent years has slowed as the need for data has diminished. However, this has not slowed advancements in cement and concrete technology, including the use of high-strength concretes. For purposes of these provisions, high-strength concrete is defined as concrete with a specified compressive strength, f'_c, of more than 10,000 psi (69 MPa).

To produce high-strength concrete, a finely divided mineral admixture must generally be used in conjunction with portland cement. Two of the most commonly used mineral admixtures are fly ash and silica fume. When these materials are used with portland cement and a high-range water-reducing admixture to yield a very low water–cement ratio concrete, it is possible to produce concrete with a compressive strength of 20,000 psi (138 MPa) or greater.

High-strength concrete has a very dense cement paste. This may adversely affect movement of water vapor or steam to the surface of a heated concrete element when it is exposed to fire. If this movement is sufficiently slowed or stopped, pressure in the pores of the cement paste can be high enough to cause spalling of the concrete. Since little research has been done in this area, application of these provisions is limited to concrete with specified compressive strengths, f'_c, of 10,000 psi (69 MPa) or less.

Analysis of the data generated in fire tests has permitted the variables that affect fire endurance to be thoroughly understood. Presentation of these data in tabular, and graphical format permits one to predict the fire resistance rating that a given assembly would achieve if subjected to the ASTM E119 test procedure. The benefits are obvious. First, one generally can calculate the rating of an assembly in less time than it would take to locate the results of a tested assembly similar to that being proposed. Second, if a fire test needs to be conducted because an identical assembly has never been tested, it would be time-consuming and would cost a considerable amount of money. Finally, because of the generic nature of concrete, a very limited number of cast-in-place concrete assemblies is found in the UL Fire Resistance Directory.[2]

Fire resistance ratings determined in accordance with these procedures are based on the fire exposure and acceptance criteria of ASTM E119 as required by most building codes. The ASTM E119 fire exposure is established by the standard time-temperature curve shown in Figure C2-1.

The conditions of acceptance imposed on walls, floors, and roofs by ASTM E119 stipulate that during the rating classification time period the assembly shall: (1) sustain the applied load (not applicable to (non–load-bearing walls); (2) prevent the passage of flame or gases hot enough to ignite cotton waste; and (3) limit the transmission of heat through the assembly such that the average temperature rise on the unexposed surface does not exceed 250 °F (139 °C) or the temperature rise at a single point does not exceed 325 °F (181 °C).

The ASTM E119 Standard also requires that a wall achieving a fire resistance rating of 1 hour or greater be subjected to the impact, erosion, and cooling effects of a hose steam test. Criteria limiting the temperature of tension reinforcement apply to all unrestrained beams, to restrained beams spaced more than 4 feet on center, and to unrestrained floor and roof

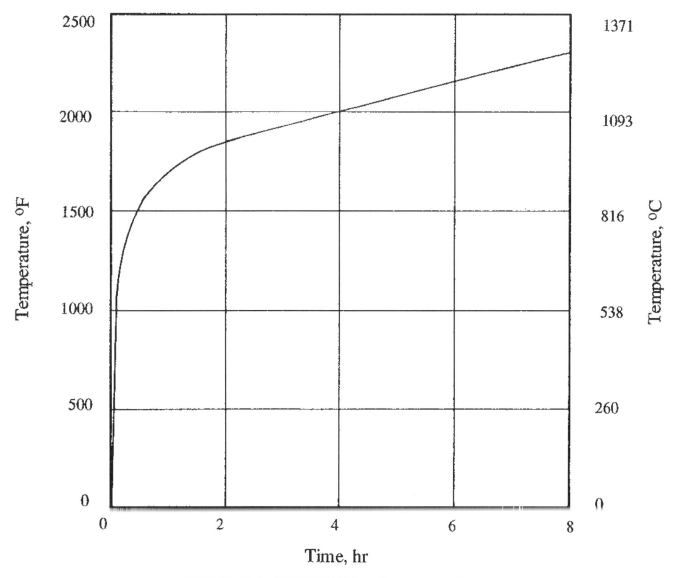

FIGURE C2-1. ASTM E119 Time-Temperature Curve

slabs having spans exceeding those tested. The only condition of acceptance for a concrete column is that it sustain the applied load.

In the case of a floor, roof, or load-bearing wall, the ASTM E119 acceptance criterion is specifically intended to assess an assembly's ability to contain a fire while continuing to support any superimposed load. For a column or beam, the criterion assesses the element's ability to support the superimposed load while being subjected to the fire exposure. For a non–load-bearing wall, the criterion assesses the wall's ability to contain a fire.

The criterion used within this Standard to assess an assembly's ability to contain a fire is the limitation

of temperature rise above ambient conditions on the unexposed surface to an average of 250 °F (139 °C).

Tests of concrete slabs have shown that aggregate type and thickness or equivalent thickness of the slab are the two variables having the most influence on the temperature rise on the unexposed surface. Therefore, adequate thickness of concrete must be provided, based on the type of aggregate used to make the concrete. Some concrete assemblies may have other materials attached to the surface of the concrete that will serve as insulation and slow the temperature rise. This Standard contains procedures for giving credit to many such commonly used materials that enhance fire resistance.

Other variables that have somewhat less effect on the fire endurance of concrete, as determined by the unexposed surface temperature rise criterion, include moisture content, air content, and maximum aggregate size. Of these, moisture content probably has the most effect. Consequently, the ASTM E119 test procedure requires that test specimens, just before fire testing, have

> a moisture condition . . . approximately representative of that likely to exist in similar construction in buildings. For purposes of standardization, this condition is to be considered as that which would be established at equilibrium resulting from drying in an ambient atmosphere of 50% relative humidity at 73 °F (23 °C).

For thicker concrete slabs it may take more than a year to reach this equilibrium moisture condition. Therefore, ASTM E119 permits the equilibrium moisture content to be that which would be obtained by drying in air at 50 to 75% relative humidity at 73 ± 5 °F (23 ± 3 °C). In addition, Appendix X4 of ASTM E119 contains procedures for adjusting a fire endurance rating determined under nonstandard moisture conditions to reflect that which would have been obtained if a standard moisture condition had existed before fire testing. The provisions in Section 2 are based on concrete conditioned to the standard moisture condition prescribed by ASTM E119, or on fire resistance ratings that have been adjusted to reflect standard conditions.

Since load-bearing assemblies must sustain the superimposed load while being subjected to the standard time-temperature fire conditions, the concrete and reinforcing steel must be able to provide the required strength at elevated temperatures. This involves providing adequate concrete cover over the steel reinforcement so that the stress induced in the reinforcement is less than its yield stress, which is commonly referred to as "yield strength." Tests of hot rolled steel reinforcing bars show that at a temperature of approximately 1,100 °F (593 °C), the yield strength of the steel is reduced to approximately 50% of its yield strength at ambient temperature conditions. Similar tests of cold drawn prestressing tendons show that at approximately 800 °F (427 °C), the tensile strength is approximately 50% of that at ambient conditions. Therefore, cover requirements of this Chapter are based on limiting the reinforcing steel and prestressing steel temperatures to 1,100 and 800 °F (593 and 427 °C), respectively. Therefore, the architect or engineer should always check concrete cover requirements for fire resistance in accordance with the provisions of this Standard.

Most fire testing on non-prestressed reinforced concrete has been performed on specimens containing uncoated steel bars. In recent years, epoxy-coated reinforcement has been widely used in bridge decks and to a lesser extent in parking structures exposed to corrosive environments. Results of fire tests reported in "Pullout Tests of Epoxy-Coated Bars at High Temperatures,"[3] "Fire Test of Concrete Slab Reinforced with Epoxy-Coated Bars,"[4] and *Fire Tests of Concrete Beams Reinforced with Epoxy-Coated Bars*[5] indicate that epoxy-coated reinforcing bars can be substituted for uncoated bars without adversely affecting the fire resistance rating of the assembly.

Several tables within the provisions address both "restrained" and "unrestrained" conditions. Restraint, in the case of assemblies tested in accordance with ASTM E119, results when expansion at supports due to the effects of heat is resisted by external forces, usually a restraining frame. When a restraining frame is not used, an unrestrained condition exists.

In the case of a concrete floor or roof slab within a typical concrete frame building, restraint is provided by the surrounding slab. As a portion of the slab is heated from below, it tries to expand; however, the expansion is resisted or restrained by the cooler concrete surrounding the area being heated. Even slabs of bays along the perimeter or at the corner of a cast-in-place concrete building are restrained, although not to the same degree as the slab in an interior bay.

A detailed discussion of restrained and unrestrained ratings is beyond the scope of this Commentary; however, suffice it to say that restraint to thermal expansion will generally enhance the fire resistance of an assembly. Cast-in-place concrete, because it is monolithic, is generally considered restrained. Precast concrete may be considered restrained or unrestrained, depending on support and end conditions and whether a structural concrete topping is provided (see Appendix A).

Guide to Use of Procedures

Use of these procedures should follow a systematic approach so that all the limiting criteria indicated above are met. The following is suggested:

General

1. Determine the required fire resistance rating.
2. Determine the type of aggregate to be used to produce concrete. (Note: More than one type of aggregate may be used to produce concrete on a project.

   For example, columns may contain normal weight aggregate concrete, whereas floor slabs may be of sand-lightweight aggregate concrete.) If the type of aggregate to be used is not known, the use of siliceous aggregate should be assumed.

Thickness of a Wall, Floor, or Roof Slab

1. Determine the thickness or equivalent thickness of a wall, floor, or roof slab.
2. Check to see if the thickness or equivalent thickness provided is equal to or greater than the required thickness from Table 2-1 or Figure 2-1, 2-2, 2-4, 2-5 (a), 2-5 (b) or the numerical solution in Section 2.3.4.2.

Cover for Steel Reinforcement in a Floor or Roof Slab or Beam

1. Determine the cover provided for reinforcement in the floor or roof slab or beam.
2. Determine the beam width and/or cross-sectional area.
3. Determine if the concrete has conventional or pre-stressed reinforcement or a combination of these. Generally, precast concrete is prestressed.
4. Determine if the floor or roof slab or beam is restrained or unrestrained in accordance with Appendix A.
5. Check to see if the cover provided is equal to or greater than that required by Table 2-5, 2-6, or 2-7 based on condition of restraint. (Note that cover requirements for non-prestressed and prestressed concrete beams are presented in different tables. Make sure the correct table is used.)

Minimum Dimension of Column and Concrete Cover for Reinforcement

1. Determine the minimum column dimension and concrete cover.
2. Check to see if the minimum dimension provided is equal to or greater than that required by Table 2-8.
3. Check to see if cover provided is equal to or greater than required by Section 2.6.1.

C2.3 Concrete Walls

Even though there have been few fire tests of concrete walls (other than concrete masonry), there have been many fire tests of concrete slabs tested as floors or roofs. Fire tests of floors or roofs are considered to be more severe than those of walls because floors and roofs must support their service (live) loads during the fire tests. In addition, most ASTM E119 fire tests of floor or roof assemblies have been conducted while the assembly was supported within a restraining frame. As concrete assemblies are heated and tend to expand, the expansion is resisted by the restraining frame. These in-plane restraining forces are usually much greater than the forces due to the superimposed load on a bearing wall. Thus floor and roof assemblies are subject to both vertical superimposed (out-of-plane) loads and horizontal restraining (in-plane) loads during fire tests. By contrast, load-bearing walls are only subjected to superimposed (in-plane) loads.

The fire endurance of masonry or concrete walls is almost always governed by the ASTM E119 criterion for temperature rise on the unexposed surface (i.e., the "heat transmission" end point). For flat concrete slabs or panels, the heat transmission fire endurance depends primarily on the aggregate type and thickness, and is essentially the same for floors as for walls. The information in Table 2-1 was derived from data in *Fire Endurance of Concrete Slabs as Influenced by Thickness, Aggregate Type, and Moisture*[6] and *Fire Resistance of Reinforced Concrete Floors.*[7]

Some building codes (e.g., The *BOCA National Building Code* and the *National Building Code of Canada*) also permit the heat transmission end point to be exceeded for both bearing and nonbearing exterior walls if the actual percentage of openings in the wall is less than permitted. In this case, the actual percentage of openings is adjusted (increased) to compensate for the additional radiation that will be emitted from the opaque portion of the wall because it is hotter than allowed by ASTM E119. The percentage of actual openings is adjusted by use of the following formula:

$$A_e = A + (A_f \times F_{eo}) \qquad (C2\text{-}1)$$

where

A_e = equivalent area or percentage of openings
A = actual area or percentage of openings
A_r = area or percentage of exterior wall surface, excluding openings, on which the temperature rise limitations of ASTM E119 are exceeded
F_{eo} = an "equivalent opening factor"

The "equivalent opening factor," F_{eo}, is determined from the following formula:

$$F_{eo} = \frac{(T_u + 459.4)^4}{(T_e + 459.4)^4} \qquad (C2\text{-}2)$$

In SI units

$$F_{eo} = \frac{(T_u + 273)^4}{(T_e + 273)^4} \qquad (C2\text{-}3)$$

where

T_u = the temperature in °F (or °C for SI units) of the unexposed wall surface at the time the required fire resistance rating is reached under ASTM E119 test conditions

T_e = 1,700 °F (927 °C) for a 1-hour fire resistance rating, 1,850 °F (1,010 °C) for a 2-hour fire resistance rating, 1,925 °F (1,052 °C) for a 3-hour fire resistance rating, or 2,000 °F (1,093 °C) for a 4-hour fire resistance rating

Since T_u values for concrete walls are not readily available for calculating F_{eo} from the above formula, Figure C2-2 may be used to determine F_{eo} for such walls for required fire resistance ratings of 1 or 2 hours.

To illustrate the use of the above, suppose that the building code requires a certain exterior wall to have a fire resistance rating of 2 hours, and the allowable percentage of openings for the particular wall is 45%. The design calls for the wall to have 40% openings. Determine the minimum thickness of a carbonate aggregate concrete wall permitted.

In this case, A_e = 45%, A = 40%, A_f = 60% (100 − 40), and F_{eo} is unknown. Therefore, the above equation must be rearranged to solve for F_{eo}.

$$F_{eo} = \frac{A_e - A}{A_f} = \frac{45 - 40}{60} = 0.083 \quad (C2\text{-}4)$$

From Figure C2-2 for carbonate aggregate concrete for 2 hours and F_{eo} of 0.083, the minimum thickness concrete wall required is 3 in. (76 mm). If the unexposed surface temperature rise limitation of ASTM E119 were not allowed to be exceeded, Table 2-1 of the Standard would require the wall thickness to be 4.6 in. (117 mm). The temperature of the unexposed surface of the 3-in. (76 mm) thick wall after 2 hours of ASTM E119 fire exposure can be calculated by rearranging Equation C2-2 to solve for T_u as shown in Equation C2-5 for U.S. customary units.

$$T_u = (F_{eo} \times (T_e + 459.4)^4)^{0.25} - 459.4 \quad (C2\text{-}5)$$

Performing this calculation for the above example, the unexposed surface temperature of the 3-in. (76-mm) thick wall after 2 hours of fire exposure is approximately 780 °F (416 °C). The temperature of a 4.6-in. (117-mm) carbonate aggregate wall after 2 hours would be approximately 325 °F (163 °C). This assumes an ambient temperature at the beginning of the fire test of 75 °F (24 °C) plus an average temperature rise of 250 °F (139 °C).

C2.3.1 Hollow-Core Panel Walls

The method for determining "equivalent thickness" of masonry units was developed because the cores in masonry units taper. The method is, of course, applicable to hollow-core precast concrete panels. However, because the cores in hollow-core panels do not taper, the equivalent thickness can be calculated by dividing the net cross-sectional area of the panel by its width.

Tests have shown that filling the cores of concrete masonry units with loose lightweight aggregates increases the fire endurance to a duration significantly longer than that of solid units of the same total thickness. Test results were reported in *Tests of the Fire*

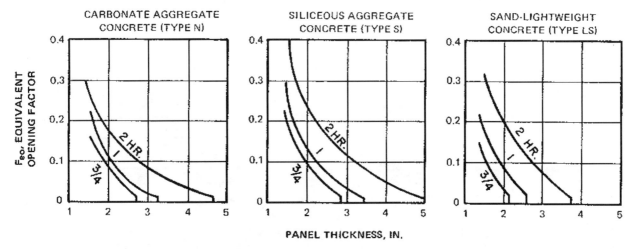

FIGURE C2-2. Equivalent Opening Factor, F_{eo}, for Concrete Walls

Resistance and Strength of Walls of Concrete Masonry Units.[8] It is reasonable to assume that the same relationship exists for walls made of hollow-core precast concrete panels.

C2.3.2 Flanged Wall Panels

Some precast concrete wall panel sections (e.g., some single-tee units) have tapered sections so the thickness varies. In fire tests it has been customary to monitor the unexposed surface temperature at a distance of two times the minimum thickness or 6 in. (152 mm), whichever is less, from the point of minimum flange thickness as shown in Figure C2-3.

C2.3.3 Ribbed or Undulating Panels

The portion of a ribbed panel that can be used in calculating the equivalent thickness, t_e, is shown in Figure C2-4.

C2.3.4 Multiple-Wythe Walls

C2.3.4.1 Graphical Solution: The graphs in Figure 2-1 were taken from *Fire Endurance of Two-Course Floors and Roofs.*[9] See Commentary Section C2.3 for a discussion on applicability of fire testing of floors to walls.

With regard to walls, some building codes permit exterior walls with specified setbacks from property lines to be unrated from the outside. In these cases, a calculation assuming the outside of the wall as the fire-exposed side is not necessary.

See Commentary Section C2.4.3 for formulas giving approximate solutions to these graphs.

C2.3.4.2 Numerical Solution: The equation in Section 2.3.4.2 is found in NBS Report BMS92, *Fire Resistance Classifications of Building Constructions.*[10] It has long been recognized that the fire resistance (based on the unexposed surface temperature rise criterion of ASTM E119) of two layered materials having different thermal conductivities will vary depending on the orientation of the two materials with respect to the fire. If the material having the lower thermal conductivity is exposed to the fire, the fire resistance will be higher than if it is oriented away from the fire. However, since walls generally are required to be fire resistance rated from both sides, the orientation resulting in the lower rating is the one that governs. Since the numerical method ignores the orientation of the layers with respect to the fire, the graphical method generally is more accurate. However, the formula has been in use for more than 50 years and is the only method avail-

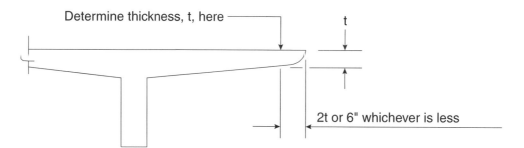

FIGURE C2-3. Flanged Wall Panels

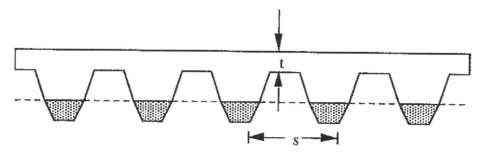

Neglect shaded area in calculation of equivalent thickness

FIGURE C2-4. Minimum Thickness for Ribbed Panels

able for calculating the fire resistance of multicourse (i.e., three or more) slabs or of assemblies incorporating air spaces.

C2.3.4.2.1 Sandwich Panels: A fire test conducted on a sandwich panel was reported in *Fire Safety Problems with Foam Plastic in the Concrete Construction Industry*.[11] The panel consisted of a 2-in. (51-mm) base slab of carbonate aggregate concrete, a 1-in. (25-mm) thickness of polystyrene insulation, and a 2-in. (51-mm) face slab of carbonate aggregate concrete. The resulting fire endurance, based on the heat transmission end point, was 120 minutes. From the equation in Section 2.3.4.2, the contribution of the 1-in. (25-mm) thickness of foam plastic polystyrene insulation was calculated to be 5 minutes.

It is likely that the contribution for a 1-in. (25-mm) thickness of foam polyurethane insulation would be somewhat greater than that for a 1-in. (25-mm) thickness of polystyrene; however, test data are not available for verification.

C2.3.4.2.2 Air Spaces: The values of $R_n^{0.59}$ for air spaces were derived from NBS Report BMS92, *Fire Resistance Classifications of Building Constructions*.[10]

C2.3.5.2 *Thickness of Insulation:* Figure 2-3 was derived from data in a report, "Fire Tests of Joints Between Precast Concrete Wall Units: Effect of Various Joint Treatments."[12]

Example: Determine the thickness of ceramic fiber blanket needed for a 2-hour fire resistance rating for joints between 5-in. (127-mm) thick precast concrete wall panels made of siliceous aggregate concrete if the maximum joint width is 3/4 in. (19 mm).

Solution: Figure 2-3 gives thicknesses of ceramic fiber blanket for 5-in. (127-mm) panels for 2-hour ratings of 0.7 in. (18 mm) for a 3/8-in. (10-mm) wide joint and 2.1 in. (53 mm) for a 1-in. (25-mm) wide joint. By direct interpolation for a 3/4-in. (19-mm) wide joint, the required thickness is 1.55 in. (39 mm).

C2.3.6 Walls with Gypsum Wallboard or Plaster Finishes

The information contained in this section is based on *Fire Endurance Tests on Unit Masonry Walls with Gypsum Wallboard*.[13]

C2.3.6.1 *Calculation for Non–Fire-Exposed Side:* The fire resistance of concrete walls generally is determined by temperature rise on the unexposed surface, i.e., the "heat transmission" end point (see Commen-

tary Section C2.3). The time required to reach the heat transmission end point (fire resistance rating) is primarily dependent on the thickness of the concrete and the type of aggregate used to make the concrete. When additional finishes are applied to the non–fire-exposed side of the wall, the time required to reach the heat transmission end point is delayed and the fire resistance rating of the wall is thus increased. The increase in rating contributed by the finish can be determined by considering the finish as adding to the thickness of concrete. However, since the finish material and concrete may have different insulating properties, the actual thickness of finish may need to be corrected to be compatible with the type of aggregate used in the concrete. The correction is made by multiplying the actual finish thickness by the factor determined from Table 2-2 and then adding the corrected thickness to the thickness of the concrete. This equivalent thickness of concrete is used to determine the fire resistance rating from Table 2-1, Figure 2-1, or Figure 2-2.

C2.3.6.2 *Calculation for Fire-Exposed Side:* When a finish is added to the fire-exposed side of a concrete wall, the finish's contribution to the total fire resistance rating is based primarily on its ability to remain in place during a fire, thus affording protection to the concrete wall. Table 2-3 lists the times that have been assigned to the finishes on the fire-exposed side of the wall. These "time assigned" values are based on results of actual fire tests. The "time assigned" values are added to the fire resistance rating of the wall alone or to the rating determined for the wall and any finish on the non–fire-exposed side.

C2.3.6.3 *Assume Each Side of the Wall Is Fire-Exposed Side:* Some building codes permit exterior walls with specified minimum setbacks from property lines to be unrated from the outside. Thus in these cases, a calculation assuming the outside of the wall as the fire-exposed side is not necessary.

C2.3.6.4 *Minimum Rating Provided by Concrete:* Where gypsum wallboard or plaster finishes are applied to one or both sides of a concrete wall, the calculated fire resistance rating for the concrete alone should not be less than one-half the required fire resistance rating. In addition, for load-bearing walls the concrete alone must provide at least twice the fire resistance as the finish on the non–fire-exposed side of the wall. This is necessary since the application of additional finishes serves to delay the heat transmission end point without adding any significant contribution to the load-carrying capability of the wall.

Example: An exterior bearing wall of a building is required to have a 2-hour fire resistance rating. The wall will be cast-in-place with siliceous aggregate concrete. It will be finished on the exterior with 5/8 in. (15.9 mm) stucco (portland cement-sand plaster) applied directly to the concrete. The interior will be finished with a 1/2-in. (12.7-mm) thickness of gypsum wallboard applied to steel furring members. What is the minimum thickness of concrete required?

First Calculation: Assume the interior to be the fire-exposed side.

1. From Table 2-3 the 1/2 in. (12.7 mm) gypsum wallboard has a "time assigned" of 15 minutes; therefore, the fire resistance rating that must be developed by the concrete and stucco on the exterior must not be less than 1 3/4 hours (2 hours minus 15 minutes).
2. From Table 2-2 the multiplying factor for portland cement–sand plaster and siliceous aggregate concrete is 1.00; therefore, the actual thickness of stucco can be added to the thickness of concrete for use in Table 2-1.
3. Since Table 2-1 does not have required thicknesses for 1 3/4 hours, direct interpolation between the values for 1 1/2 and 2 hours is acceptable. The interpolation results in a required thickness of 4.65 in. (118 mm) of concrete and stucco. Since the stucco is 5/8 in. (15.9 mm) thick, the concrete must be at least 4.02 in. (102 mm) (i.e., 4.65–0.63).

Second Calculation: Assume the exterior to be the fire-exposed side.

1. From Table 2-2 the multiplying factor for gypsum wallboard and siliceous aggregate concrete is 3.00; therefore, the corrected thickness for 1/2 in. (12.7 mm) of gypsum wallboard is 1.5 in. (38 mm) (i.e., 3.00 × 0.5).
2. Footnote 1 to Table 2-3 allows 5/8 in. (15.9 mm) of stucco applied directly to the concrete to be added to the actual thickness of concrete rather than establishing a "time assigned" value.
3. Table 2-1 requires 5.0 in. (127 mm) of siliceous aggregate concrete for a 2-hour fire resistance rating. Therefore, the actual thickness of concrete required is 2.87 in. (73 mm) (i.e., 5.00–1.5–0.63).
4. Since the thickness of concrete required when assuming the interior side to be the fire-exposed side is greater [i.e., 4.02 in. (102 mm)], this is the minimum concrete thickness allowed to achieve a 2-hour fire resistance rating.

5. Section 2.3.6.4 requires that the concrete alone provide not less than one-half the total required rating.

Thus, the concrete must provide at least a 1-hour rating. From Table 2-1 it can be seen that only 3.5 in. (89 mm) of siliceous aggregate concrete is required for 1 hour, whereas 4.02 in. (102 mm) will be provided.

C2.3.6.5.1 Furring Members: Wood furring members, although combustible, do not adversely affect the fire resistance rating of the wall. This is due to the fact that they are protected from the fire by the finish for a period of time deemed to be equal to the "time assigned" value of the finish (see Table 2-3).

C2.4 Concrete Floor and Roof Slabs

The fire test criteria for temperature rise of the unexposed surface and the ability to resist superimposed loads (heat transmission and structural criteria, respectively) must both be considered in determining the fire resistance of floors and roofs. Section 2.4 deals with heat transmission and Section 2.5 addresses the structural criterion.

The criterion limiting the average temperature rise to 250 °F (139 °C) and the maximum rise at one point to 325 °F (181 °C) is often referred to as the "heat transmission end point." For concrete slabs, the heat transmission end point is mainly a function of slab thickness (or equivalent thickness) and aggregate type. Other factors that affect heat transmission to a lesser degree are moisture content of the concrete, maximum aggregate size, mortar content, and air content. Items that have very little effect on heat transmission are cement content, concrete strength, and amount and location of reinforcement, provided these items are within the normal range of usage. The values in Table 2-1 apply to concrete slabs reinforced with bars or welded wire fabric as well as to prestressed slabs.

The information in Table 2-1 was derived from data in *Fire Endurance of Concrete Slabs as Influenced by Thickness, Aggregate Type, and Moisture* and *Fire Resistance of Reinforced Concrete Floors.*[7]

C2.4.2 Joints in Precast Slabs

Based on data developed by Underwriters Laboratories, the *Fire Resistance Directory*[2] indicates that where no concrete topping is used over precast concrete floors, joints must be grouted or the fire resistance of the slab maintained by another method that has been tested and approved by the authority having jurisdiction. However, if a concrete topping at least 1 in. thick is used, the joints need not be grouted or otherwise protected.

C2.4.3 Two-Course Floors and Roofs

Figure 2-4 was derived from ASTM E119 testing conducted by Portland Cement Association and presented in *Fire Endurance of Two-Course Floors and Roofs.*[9] The following formulas are approximate representations of the curves shown in the graphs and were first presented in "Calculations of the Fire Resistance of Composite Concrete Floor and Roof Slabs."[14]

Where the base slab or side exposed to the fire is siliceous or carbonate aggregate concrete,

$$R = 0.057(2t^2 - dt + 6/t) \qquad (C2\text{-}6)$$

In SI units

$$R = 0.00018t^2 - 0.00009\,dt + 8.7/t \quad (C2\text{-}7)$$

and where the base slab or side exposed to the fire is sand-lightweight or lightweight aggregate concrete,

$$R = 0.063\,(t^2 + 2dt - d^2 + 4/t) \qquad (C2\text{-}8)$$

In SI units

$$R = 0.0001t^2 + 0.0002dt - 0.0001d^2 + 6.4/t \quad (C2\text{-}9)$$

where

R = fire resistance of slab in hours
t = total thickness of slab in in. (mm)
d = thickness of base slab in in. (mm)
$t - d \geq 1$ in. (25 mm).

C2.4.4 Insulated Roofs

Figures 2-5 (a) and 2-5 (b) were derived from *Fire Resistance of Lightweight Insulating Concretes*[15] and *Fire Endurance of Two-Course Floors and Roofs,*[9] respectively. The provisions allowing 10 minutes to be added to the fire resistance determined from 2-5 (a) to account for the contribution of a standard 3-ply built-up roof is conservative. It is based on a comparison of fire test results of otherwise similar assemblies with and without a built-up roof and reported in *Fire Endurance of Two-Course Floors and Roofs.*[9]

C2.5 Concrete Cover over Reinforcement

ASTM E119 differentiates between restrained and unrestrained floor, roof and beam fire endurance classifications. "Restraint" in this case means the restriction of thermal expansion imposed by a restraining frame during a fire test (see Commentary, Section 2.1). Appendix A, which is taken from Appendix X3 of ASTM E119,[1] gives guidelines for determining whether a

floor, roof, or beam can be considered restrained. Appendix A indicates that some concrete slabs supported by walls are considered unrestrained. Note "a" to the table in the Appendix indicates that if certain conditions are satisfied, these members can be considered restrained. To achieve restraint, the wall, spandrel beam, or other member providing resistance must be specifically designed to withstand the thermal expansion of the beam or slab. Guidance for determining thermal expansion of beam and slab systems and for designing restraining members can be found in *Reinforced Concrete Fire Resistance*[16] and *PCI Design for Fire Resistance of Precast Prestressed Concrete.*[17]

In addition, ASTM E119 gives different structural criteria for restrained and unrestrained assemblies. For unrestrained concrete floors, roofs, or beams tested in a restrained condition, the fire test end point occurs when the average temperature of the tensile reinforcement reaches 1,100 °F (593 °C) for hot-rolled reinforcing bars or 800 °F (427 °C) for cold-drawn prestressing steel.

For restrained beams spaced more than 4 ft (1,219 mm) on center, the temperatures noted above must not be exceeded during the first one-half of the classification period or 1 hour, whichever is greater. For example, if a restrained beam is required to have a 2-hour fire resistance rating, the average temperature of the steel tensile reinforcement is not permitted to exceed the limiting value during the first hour of the fire test. The temperature limits do not apply to restrained beams spaced 4 ft (1,219 mm) or less on center or to restrained slabs.

C2.5.1 Cover for Slab Reinforcement

The temperature of the tensile reinforcement in slabs depends on the thickness of the concrete cover and aggregate type. For unrestrained slabs, the cover thickness shown in Table 2-5 for nonprestressed and prestressed reinforcement are those needed to keep the tensile reinforcement below 1,100 and 800 °F (593 and 427 °C), respectively. For restrained slabs, the temperature of the tensile reinforcement is not critical and thus, a cover of 3/4 in. (19 mm) is specified. Three-fourths (3/4) in. is the minimum permitted by ACI 318 *Building Code Requirements for Structural Concrete.*[18] The cover values for unrestrained slabs were derived from *Fire Endurance of Concrete Slabs as Influenced by Thickness, Aggregate Type, and Moisture.*[6]

C2.5.2 Cover for Non-Prestressed Reinforcement in Beams

The temperature of reinforcement in concrete beams depends on beam width and cover thickness.

For temperatures above approximately 1,000 °F (538 °C), the effect of aggregate type is minimal as shown in "Fire Endurance of Concrete Slabs as Influenced by Thickness, Aggregate Type, and Moisture"[6]; however, for lower temperatures, differences in aggregate types are more pronounced.

For non-prestressed reinforced concrete beams, the critical steel temperature of ASTM E119 is 1,100 °F (593 °C), so the effect of aggregate type is minimal as indicated above. The data in Table 2-6 were derived from fire tests of beam specimens that ranged in width from 2 to 24 in. (51 to 610 mm). Other variables included aggregate type and amount of reinforcement. The tests were conducted at the Portland Cement Association and reported in "Measured Temperatures in Concrete Beams Exposed to Fires."[19] Results of fire tests conducted at Underwriters Laboratories were also analyzed.

C2.5.3 Cover for Prestressed Reinforcement in Beams

As indicated in Commentary Section C2.5, ASTM E119 limits cold-drawn prestressing steel temperatures to 800 °F (427 °C). Thus, aggregate used in the concrete must be considered for prestressed concrete beams.

Cover requirements found in Table 2-7 (a) were adapted from a similar table in Appendix D of the National Building Code of Canada 1995.[20] The original research for the table was contained in "Fire Endurance of Concrete Assemblies,"[21] a compilation of published information on fire endurance of a variety of concrete walls, floors, roofs, columns, and beams. These provisions can be used for beams of any width; however, they are generally applied to thin beams cast monolithically with precast slab systems.

The data in Table 2-7 (b), which are applicable to beam widths equal to or greater than 8 in. (203 mm), were derived from fire tests of beam specimens that ranged in width from 2 to 24 in. (51 to 610 mm). Other variables included aggregate type and amount of reinforcement. The tests were conducted at the Portland Cement Association and reported in "Measured Temperatures in Concrete Beams Exposed to Fires."[19] Results of fire tests conducted at Underwriters Laboratories were also analyzed.

C2.6 Reinforced Concrete Columns

In the past, most code provisions in the United States for reinforced concrete columns have been based on two reports, "Fire Tests of Building Columns"[22] and "Fire Resistance of Concrete Columns."[23] Sizes of columns tested were 12, 16, and 18 in. (305, 406, and 457 mm). Nearly all of the columns withstood 4 hours of fire test exposure conducted essentially in accordance with ASTM E119. Most of the tests were stopped after 4 hours, but some were continued for 8 hours. The shortest duration was 3 hours. At the time of the tests, few if any concrete columns were smaller than 12 in. (305 mm), but smaller columns have been in use since that time. Fire tests conducted in Europe, notably Great Britain, under essentially the same fire test procedure as ASTM E119, have shown that smaller columns have somewhat less fire resistance. Many of the columns tested are described in "Rehabilitation Guidelines 1980; Vol. 8—Guideline on Fire Ratings of Archaic Materials and Assemblies."[24]

The most recent fire testing of loaded concrete columns was a joint venture between the Portland Cement Association and the National Research Council of Canada and reported in "Fire Resistance of Reinforced Concrete Columns,"[25] and "Fire Resistance of Reinforced Concrete Columns—Test Results."[26] The data developed in this program were used to update minimum column dimensions shown in Table 2-8.

C2.6.1 Minimum Cover for Reinforcement

Cover requirements are based on provisions contained in Appendix D of the National Building Code of Canada 1995.[20] The requirements are based on calculation methods described in "Further Studies of the Fire Resistance of Reinforced Concrete Columns"[27] and "Fire Performance of Reinforced Concrete Columns."[28] They were validated using the results of more than 40 fire tests reported in "Fire Resistance of Reinforced Concrete Columns—Test Results."[26]

This will permit slightly less cover where required fire ratings are less then 1 1/2 hours than has traditionally been required in the United States. However, it should be pointed out that for cast-in-place concrete, ACI 318 "Building Code Requirements for Structural Concrete"[18] requires 1 1/2 in. (38 mm) of concrete cover to column ties or spirals. Therefore, if the tie or spiral is #3 or #4 bars, the cover to the main longitudinal reinforcement will be 1 7/8 or 2 in. (48 or 51 mm), respectively. For precast concrete columns, ACI 318 requires the cover for the longitudinal reinforcing steel to be not less than the diameter of the steel or 5/8-in. (16 mm), whichever is larger; and 3/8-in. (10 mm) cover for ties and spirals. Columns fire tested at NRCC and Reported in "Fire Resistance of Reinforced Concrete Columns"[25] had 1-7/8 in. (48 mm) cover.

C2.6.2 Columns Built into Walls

The provisions of this Section are based on Appendix D of the National Building Code of Canada 1995.[20] They require that the wall into which the column is built have an equal or higher rating than required for the column and that openings in the wall be protected. This ensures that the column will be exposed to fire on one side only, allowing the column to be considered a wall for fire resistance purposes.

REFERENCES FOR COMMENTARY 2

1. American Society for Testing and Materials. *Standard Methods of Fire Tests of Building Construction and Materials,* ASTM Designation E119-88, West Conchohoken, Pa., 1988.
2. Underwriters Laboratories Inc. *Fire Resistance Directory,* Northbrook, Ill., 1992.
3. Lin, T.D., R.I. Zwiers, S.T. Shirley, and R.G. Burg. "Pullout Tests of Epoxy-Coated Bars at High Temperatures," *Materials Journal,* American Concrete Institute, Detroit, Mich., November–December 1988.
4. Lin, T.D., R.I. Zwiers, S.T. Shirley, and R.G. Burg. "Fire Test of Concrete Slab Reinforced with Epoxy-Coated Bars," *Structural Journal,* American Concrete Institute, Detroit, Mich., March–April 1989.
5. Lin, T.D. and R.G. Burg. *Fire Tests of Concrete Beams Reinforced with Epoxy-Coated Bars,* RP321, Portland Cement Association, Skokie, Ill., 1994.
6. Abrams, M.S. and A.H. Gustaferro. *Fire Endurance of Concrete Slabs as Influenced by Thickness, Aggregate Type, and Moisture,* PCA Research and Development Laboratories, 10(2), May 1968. Also, PCA Research Department Bulletin 223.
7. Thompson, John P. *Fire Resistance of Reinforced Concrete Floors,* Portland Cement Association EB065. Skokie, Ill., 1963.
8. Menzel, Carl A. *Tests of the Fire Resistance and Strength of Walls of Concrete Masonry Units,* Portland Cement Association, Skokie, Ill., January 1934.
9. Abrams, M.S. and A.H. Gustaferro. *Fire Endurance of Two-Course Floors and Roofs,* Portland Cement Association Research and Development Bulletin RD048, Skokie, Ill., Also, *J. Am. Concrete Inst.,* February 1969.
10. National Bureau of Standards. *Fire Resistance Classifications of Building Constructions,* Report BMS 92, National Bureau of Standards, Washington D.C., 1942.
11. Gustaferro, Armand H. *Fire Safety Problems with Foam Plastic in the Concrete Construction Industry,* Precast/Prestressed Institute, Chicago, Ill.
12. "Fire Tests of Joints Between Precast Concrete Wall Units: Effect of Various Joint Treatments," *PCI J.,* September–October, 1975.
13. Allen, L.W., M. Galbreath, and W.W. Stanzak. *Fire Endurance Tests on Unit Masonry Walls with Gypsum Wallboard* (NRCC 13901). Division of Building Research, National Research Council of Canada.
14. Lie, T.T. "Calculations of the Fire Resistance of Composite Concrete Floor and Roof Slabs," *Fire Technology,* 14(1), 1978.
15. Abrams, M.S. and A.H. Gustaferro. *Fire Resistance of Lightweight Insulating Concretes,* Research and Development Bulletin RD004, Portland Cement Association, Skokie, Ill., 1970.
16. Concrete Reinforcing Steel Institute. *Reinforced Concrete Fire Resistance,* Concrete Reinforcing Steel Institute, Chicago, Ill., 1980.
17. Gustaferro, Armand H. and Leslie D. Martin. *PCI Design for Fire Resistance of Precast Prestressed Concrete,* Prestressed Concrete Institute, Chicago, Ill., 1977.
18. American Concrete Institute. *Building Code Requirements for Structural Concrete,* ACI 318-95, American Concrete Institute, Detroit, Mich., 1995.
19. Lin, T.D. *Measured Temperatures in Concrete Beams Exposed to Fires,* Construction Technology Laboratories, Skokie, Ill., 1985.
20. National Research Council of Canada. *National Building Code of Canada 1995,* NRCC No. 38726, Ottawa, Canada, 1995.
21. Galbreath, Murdoch. *Fire Endurance of Concrete Assemblies.* Divisions of Building Research, DBR Technical Paper No. 235, National Research Council of Canada, Ottawa, Canada, 1966. (A compilation of published information on fire endurance of a variety of concrete walls, floors, roofs, columns, and beams.)
22. Associated Factory Mutual Insurance Companies. *Fire Test of Building Columns,* Associated Factory Mutual Insurance Companies, Johnston, R.I., 1921.
23. Hull and Ingberg. *Fire Resistance of Concrete Columns,* Technological Papers of the (National) Bureau of Standards, No. 271, February 24, 1925.

24. U.S. Department of Housing and Urban Development. *Rehabilitation Guidelines 1980; Vol. 8—Guideline on Fire Ratings of Archaic Materials and Assemblies,* U.S. Department of Housing and Urban Development, Washington D.C., 1981.

25. Lie, T.T., et al. *Fire Resistance of Reinforced Concrete Columns,* Division of Building Research, DBR Paper No. 1167, National Research Council of Canada, Ottawa, Canada, February, 1984. Also, NRCC 23065.

26. Lie, T.T. and J.L. Woolerton. *Fire Resistance of Reinforced Concrete Columns—Test Results,* Institute for Research in Construction, Internal Report No. 569, National Research Council of Canada, Ottawa, Canada, 1988.

27. Allen, D.E. and T.T. Lie. *Further Studies of the Fire Resistance of Reinforced Concrete Columns,* Division of Building Research, DBR Technical Paper No. 416, National Research Council of Canada, Ottawa, Canada, 1974.

28. Lin, T.D. and T.T. Lie. "Fire Performance of Reinforced Concrete Columns," *Fire Safety: Science and Engineering,* American Society for Testing and Materials, ASTM, STP 882, Philadelphia, Pa., 1985. Also, Division of Building Research, DBR Paper No. 1352, National Research Council of Canada, Ottawa, Canada. Also, NRCC 25351.

Commentary for 3: Standard Methods for Determining the Fire Resistance of Timber and Wood Structural Elements

C3.1.2 Dimensions and Metric Conversion

Tables with nominal and actual (dressed and dry) dimensions of commonly used sections of sawn lumber and glued laminated timber can be found in the Supplement to the 1991 Edition of the *National Design Specification for Wood Construction.*[10]

C3.3 Design of Fire-Resistive Exposed Wood Members

Exposed wood beams and columns have long been recognized for their ability to maintain structural integrity while exposed to fire. Early mill construction from the 19th century utilized massive timbers to carry large loads and to resist structural failure from fire. It is the mass of these timbers that enhances the fire performance of heavy timber beams and columns due to charring effect. During ASTM E119 fire exposure, wood will char at a rate of approximately 1/30 of an in. per minute (0.85 mm/min) for the first 15–20 minutes of exposure. After this initial time period, the charring rate decreases to 1/40 of an in. per minute (0.64

mm/min). This slower rate is attributed to the insulative effect of the ever-deepening char layer. Once a layer of char has formed on the surface of the timber member, further charring of the cross section proceeds slowly. This Chapter describes a mathematical model that allows design of wood members to comply with fire-resistive requirements.

Basis of Procedure: This empirically based mathematical model generates a conservative design for the minimum dimensions of a wood beam or column. The model accounts for the cross-sectional dimension change due to charring of the beam or column after any given period of fire exposure. The model also accounts for the effects of elevated temperatures on the load-carrying capability of the residual section. Use of this procedure will provide the required minimum dimensions to safely carry loads over a given time period under ASTM E119 fire exposure.

C3.3.1 Analytical Method for Exposed Wood Members

The analytical method for determining fire resistance ratings of up to 1 hour is based on a mathematical model validated by a series of tests, including ASTM E119 tests. Development of fire resistance rating greater than 1 hour for wood framing members must be based on actual test results.

C3.3.2.2.1 Glued Laminated Timber Beams: The outer tension lamination is the most critical part of a glued laminated beam, yet when it is directly exposed to a fire, it will be almost completely consumed at the end of 1 hour. For this reason, an extra tension lamination is to be used. This is accomplished in all bending combinations by adding an extra outer tension lamination and removing one core lamination.

C3.3.2.4 Connectors and Fasteners: The connectors and fasteners are important elements in the performance of wooden beams and columns when exposed to fire. Exposed metal connectors and fasteners must be shielded from direct exposure to the fire. This is accomplished by applying a protective membrane on top of an exposed connector or by embedding the connector within the cross section of the beam or column.

C3.4 Component Additive Method for Calculating and Demonstrating Assembly Fire Endurance

The original methodology for calculating fire endurance ratings of assemblies by CAM was developed in the early 1960s by the Fire Test Board of the National Research Council of Canada. The methodology resulted from their detailed review of 135 standard fire test reports on wood stud walls, 73 test reports on wood-joist floor/ceiling assemblies, and the "Ten Rules of Fire Endurance Rating."[3] Review of the fire tests provided assigned time values for contribution to fire endurance ratings for each separate component of an assembly. The "Ten Rules" provided a method for combining the individual contributions to obtain the fire endurance rating of the assembly.

The times assigned to various wood-frame assembly components were determined from a detailed review of more than 200 standard fire test reports, using the "Ten Rules of Fire Endurance Rating" developed by Harmathy. These rules with a brief explanation are set forth below:

Rule 1 *The "thermal" fire endurance of a construction consisting of a number of parallel layers is greater than the sum of the "thermal" fire endurance characteristics of the individual layers when exposed separately to fire.*

Where two layers of panel materials, such as gypsum board or plywood, are fastened to studs or joists adequately, their combined effect is greater than the sum of their individual contributions to the fire endurance rating of the assembly. For example, the fire endurance time assigned to 1/2 in. (12.7 mm) gypsum board is 15 minutes (see Table 3-1). Two layers of 1/2 in. (12.7 mm) gypsum board have an endurance rating greater than 15 + 15 = 30 minutes.

Rule 2 *The fire endurance of a construction does not decrease with the addition of further layers.*

This is almost the converse of Rule 1. It says that any additional layers of wallboard or other panel materials will add to fire endurance no matter how many layers are added.

Rule 3 *The fire endurance of constructions containing continuous air gaps or cavities is greater than the fire endurance of similar constructions of the same weight, but containing no air gaps or cavities.*

Wall and ceiling cavities formed by studs and joists protected and encased by wall coverings add to the fire endurance of these assemblies.

Rule 4 *The farther an air gap or cavity is located from the exposed surface, the more beneficial its effect on the fire endurance.*

In cases where cavities are formed by joists or studs and protected by 2-in.-thick (51 mm) panel materials against fire exposure, the beneficial effect of such

air cavities is greater than if the protection is only 1/2 in. thick (12.7 mm) (see Rule 7).

Rule 5 *The fire endurance of an assembly cannot be increased by increasing the thickness of completely enclosed air layer.*

Increasing stud or joist depths from 4 in. (89 mm) to 6 in. (140 mm), or even to 12 in. (286 mm), does not increase the level of fire endurance.

Rule 6 *Layers of materials of low thermal conductivity are better utilized on the side of the construction on which fire is more likely to happen.*

A building material made of wood fiber is more effective against thermal transfer than is a material having relatively high thermal conductivity, such as metal. Wood will be more effective in protecting against excessive rise in temperature on the opposite face of assemblies. This temperature rise can lead to failure under test acceptance criteria (see Rule 7).

Rule 7 *The fire endurance of asymmetrical constructions depends on the direction of heat flow.*

Walls that do not have the same panel materials on both faces will demonstrate different fire endurance ratings depending on which side is exposed to fire. This rule results as a consequence of Rules 4 and 6, which point out the importance of the location of air gaps or cavities and of the sequence of different layers of solids.

Rule 8 *The presence of moisture, if it does not result in explosive spalling, increases the fire endurance.*

Materials having a 15% moisture content will have greater "thermal" fire endurance than those having 4% moisture content at the time of fire exposure.

Rule 9 *Load-supporting elements, such as beams, girders, and joists, yield higher fire endurance when subject to fire endurance tests as parts of floor, roof, or ceiling assemblies than they would when tested separately.*

A wood joist performs better when it is incorporated in a floor/ceiling assembly than tested by itself under the same load.

Rule 10 *The load-supporting elements (beams, girders, joists, etc.) of a floor, roof, or ceiling assembly can be replaced by such other load-supporting elements that, when tested* *separately, yielded fire endurance not less than that of the assembly.*

A joist in a floor assembly may be replaced by another type of joist having a fire endurance rating not less than that of the assembly.

C3.4.1.1 Component Times: The times assigned to protective wall and ceiling coverings are given in Table 3-1. These times are based on the ability of the membrane to remain in place during fire tests. This "assigned time" should not be confused with the "finish" rating of the membrane. A "finish rating" is the time at which the wood stud or joist reaches an average temperature rise of 250 °F (140 °C), or an individual temperature rise of 325 °F (180 °C) above ambient on the plane of the wood nearest the fire. As shown in Table 3-1, some pairs of membranes have been tested resulting in assigned times greater than the sum of the assigned times of the individual membranes.

The times assigned to wood studs and joists were determined based on the time it takes for the framing members to fail after failure of the protective membrane. The fire endurance time assigned to framing members is given in Table 3-2. These times are based on the ability of framing members to provide structural support when subjected to the ASTM E119 fire endurance test without benefit of a protective membrane. These time values are in part the result of full-scale tests of unprotected wood studs and floor joist where the structural elements were loaded to design capacity. They apply to all framing members and do not increase if, for example, 2 in. × 6 in. (38 mm × 140 mm) studs are used rather than 2 in. × 4 in. (38 mm × 89 mm) studs.

Additional fire endurance can be provided to wall assemblies by the use of high-density rockwool or paper or foil-faced glass fiber insulation batts. The time assigned to each type of insulation as contributing additional fire endurance to the assembly is presented in Table 3-3.

C3.4.1.4 Floor/Ceiling and Roof/Ceiling Assemblies: The arbitrary addition of insulation in a floor/ceiling assembly that has not been methodically analyzed or tested with the insulation as part of the assembly, could reduce the fire resistance rating of the assembly.

EXAMPLES

C3.3.2.2 Beams

Example 1: *The structural members of a 2-story office building is required to have a 1-hour fire resistance. The building is 35 ft. (10.67 m) wide and 60 ft (18.29 m) in length. Repetitive member framing spans from*

the exterior walls to glued laminated beam at the center of the building. Two simply supported beams, each 30 ft (9.14 m) in length, bear on the exterior walls and a column at the center of the building. The beam is of greater depth than the floor/ceiling system; therefore a portion of the beam will be left exposed beneath the rated floor/ceiling assembly. Determine the fire endurance rating of the beam selected to carry the imposed loads. Extreme fiber in bending F_b is given as 2,400 psi (16.55 MPa), and the species is Douglas fir. The live and dead loads are 50 lb/ft² (2,394 N/m²) and 15 lb/ft² (718 N/m²), respectively. The load duration factor, C_D, is taken equal to 1.0.

Solution:

$$w = 65 \text{ lb/ft}^2 \times 17'\text{-}6''$$
$$= 1{,}137.5 \text{ lb/ft (16.6 kN/m)}$$

$$\text{Moment} = wl^2 \div 8 = 1.536 \times 10^6 \text{ in.-lb (174 kN-m)}$$

$$S_{required} = 1{,}536{,}000 \div 2{,}400$$
$$= 640 \text{ in.}^3 \ (10.5 \times 10^6 \text{ mm}^3)$$

Try a beam with actual dimensions 8 3/4″ × 24″;

$$S_{actual} = 840 \text{ in.}^3 \ (13.8 \times 10^6 \text{ mm}^3)$$

Check volume factor (Section 5.3.2 of the NDS)

$$C_V = 0.85$$
$$S_{required}/C_v = 753 \text{ in.}^3 < 840 \text{ in.}^3 \therefore \text{ O.K.}$$

Calculate the load ratio:

$$r = 753 \text{ in.}^3 \div 840 \text{ in.}^3 = 90\%$$

Calculate the load factor from Section 3.2.2.1.2:

$$z = 0.7 + 30/r = 0.7 + 30/90 = 1.033$$

Calculate t from Section 3.2.2.2, Eq. (3.4):

$$t = \gamma z b \ [4\text{-}(b/d)]$$
$$= 2.54(1.033)(8.75)[4\text{-}(8.75/24)]$$

time to failure = 83 minutes > 60 minutes

Therefore the assembly will provide a 1-hour fire resistance. Eliminate one core laminate and add one outer tension zone laminate to the bottom of the beam.

C3.3.2.3 Columns

Example 1: *The column supporting the glued laminated beams in the previous example is to be left ex-*

posed. A Douglas fir glued laminated column with an actual section of 8 3/4 in. × 9 in. (222 mm × 229 mm) has been specified. The column length is 8 ft (2.44 m). Determine the fire endurance rating of the column selected to carry the imposed loads. The compressive strength and stiffness are $F_c = 1,900$ psi (13.1 MPa), and $E = 1,700,000$ psi (11,721 MPa), respectively.

Solution:

$$\text{Area} = 78.75 \text{ in.}^2 (50.8 \times 10^3 \text{ mm}^2)$$

$$\text{Column load} = 65 \text{ lb/ft}^2 \times 35 \text{ ft} \div 2 \times 60 \text{ ft} \div 2$$

$$= 34,125 \text{ lbs (152 kN)}$$

Calculate the required compressive stress:

$$F_{c, required} = 34,125 \text{ lb} \div 78.75 \text{ in.}^2$$

$$= 433 \text{ psi (3 MPa)}$$

Determine the effective length factor from Figure 3-1:

$$K_e = 1.0$$

Calculate the effective length:

$$l_e = 1 = 8'(12) = 96 \text{ in. (2.44 m)}$$

Calculate the column stability factor (from NDS, Eq. 3.7-1):

$$C_p = 0.957$$

Calculated adjusted Fc′:

$$F_c' = 1,900 \text{ psi} \times 0.957 = 1,819 \text{ psi (12.54 MPa)}$$

Calculate load factor from Section 3.2.2.1.1:

$$\text{Load ratio,} \quad r = 433 \div 1,819 = 0.24 = 24\%$$

Calculate slenderness ratio:

$$K_e l_e / d = 1 \times 96 \text{ in.} \div 8.75 \text{ in.} = 10.97 \therefore z = 1.5$$

Calculate the time to failure from Section 3.2.2.3, Eq. (3.5):

$$t = \gamma z d \, [3-(d/b)]$$
$$= 2.54(1.50)(8.75)[3-(8.75/9)]$$
$$= 67.6 \text{ min}$$

Therefore the assembly will provide a 1-hour fire resistance.

C3.4 Component Additive Method for Calculating and Demonstrating Assembly Fire Endurance

Example 1: *Determine the fire endurance rating of a wall assembly having one layer of 5/8 in. (15.9 mm) type X gypsum board attached to wood studs on the fire-exposed side as shown in Figure C3-1.*

Solution: Table 3-1 shows that 5/8 in. (15.9 mm) Type X gypsum board has an assigned time of 40 minutes. Table 3-2 shows that wood studs spaced 16 in. (406 mm) on center have a time of 20 minutes. Summing the two components results in a fire endurance rating of 60 minutes.

If the wall is assumed to be exposed to fire from both sides (e.g., for interior fire rated walls), each surface of the framing member would be required to be

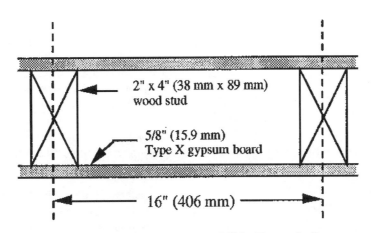

FIGURE C3-1. Fire-Exposed Side (Example 1)

fire protected with at least 40 minutes of membrane coverings in Table 3-1. If the proposed wall is assumed to be exposed to fire from the interior only, a total contribution of 40 minutes from the interior membrane coverings from Table 3-1 is required. However, in this case, the exterior side must be protected in accordance with Table 3-4, or any membrane that is assigned a time of at least 15 minutes as listed in Table 3-1.

If the wall cavities between studs had been filled with rockwool insulation adding 15 minutes of fire endurance, as noted in Table 4-3, the 5/8 in. (15.9 mm) Type X gypsum board could be replaced by 1/2 in. (12.7 mm) Type X gypsum board. Thus, adding the fire endurance contribution times for the 1/2 in. (12.7 mm) Type X gypsum board, wood studs and insulation (24 minutes + 20 minutes + 15 minutes), the resultant fire endurance rating for the wall would also equal 60 minutes.

Example 2: *Determine the fire endurance rating of a floor/ceiling assembly having wood joists spaced 16 in. (406 mm) and protected on the bottom side (ceiling side) with two layers of 1/2 in. (12.7 mm) Type X gypsum board and having a 1/2 in. (12.7 mm) plywood subfloor on the upper side (floor side) as shown in Figure C3-2.*

Solution: Table 3-1 shows that the assigned time for each layer of 1/2 in. (12.7 mm) Type X gypsum board is 25 minutes. The time assigned for wood joists, as shown in Table 3-2, is 10 minutes. Adding the assigned time of two layers of gypsum board and wood joists, a fire endurance rating of 60 minutes is calculated.

Example 3: *A private residence is being changed to an office. The load-bearing exterior walls of the residence (see Figure C3-3) consist of 2 × 4 in. (38 mm × 89*

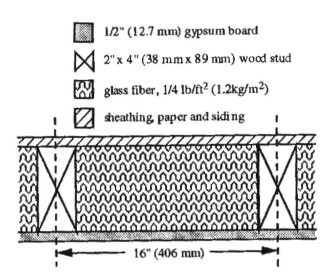

FIGURE C3-3. **Fire-Exposed Side (Example 3)**

mm) studs spaced 16 in. (406 mm) on center, 1/2 in. (12.7 mm) gypsum board on the inside, and 5/16 in. (7.9 mm) exterior grade plywood, sheathing paper and 1/4 in. (6.4 mm) hardboard siding on the outside. The cavities between the wood studs are filled with 1/4 lb/ft² (1.2 kg/m²) glass fiber batts. The code requires the exterior wall of the structure to be upgraded to 1-hour fire endurance with fire exposure from the inside only. What modification can be made to comply with code requirement?

Solution: Table 3-1 shows that 1/2 in. (12.7 mm) gypsum board has a contribution to the assembly fire rating of 15 minutes. According to Table 3-2, the studs have an assigned time of 20 minutes. According to Table 3-3, the glass fiber does not contribute to the fire endurance of a load-bearing wall. Thus, the fire endurance rating of the exterior wall of the residence equals 35 minutes. In order to upgrade the wall to 1 hour, a protective membrane shall be added on the inside, contributing 25 minutes or more to the assembly rating. For example, a 1/2 in. (12.7 mm) Type X gypsum board adds 25 minutes according to Table 3-1, leading to a total of 60 minutes.

REFERENCES FOR COMMENTARY 3

1. Galbreath, M. "Estimating the Fire Endurance of Exterior Stud Walls," Technical Paper No. 335, National Research Council of Canada, Division of Building Research, Ottawa, Canada, 1961.

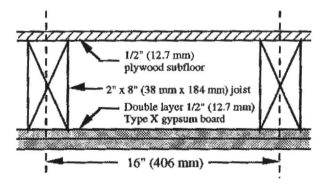

FIGURE C3-2. **Fire-Exposed Side (Example 2)**

2. Galbreath, M. "Fire Performance Rating," Canada Building Digest No. 71, National Research Council of Canada, Division of Building Research, Ottawa, Canada, 1965.

3. Harmathy, T.Z. "Ten Rules of Fire Endurance Rating," *Fire Technology,* 1(2), 1965, pp. 93–102.

4. Galbreath, M. "Fire Endurance of Light Framed and Miscellaneous Assemblies," Technical Paper No. 222, National Research Council of Canada, Division of Building Research, Ottawa, Canada, 1966.

5. Hall, G.S. "Fire Resistance Tests of Laminated Timber Beams," Research Report WT/RR/1, Timber Research and Development Association, High Wycombe, England, 1968.

6. Hall, G.S., et al. "Fire Performance of Timber," Timber Research and Development Association, High Wycombe, England, 1974.

7. Lie, T.T. "A Method for Assessing the Fire Resistance of Laminated Timber Beams and Columns," *Can. J. Civ. Eng.,* 4(2), 1977, pp. 161–169.

8. Schaffer, E.L. "State of Structural Timber Fire Endurance," *Wood and Fiber Sci.* 9(2), 1977, pp. 145–170.

9. Schaffer, E.L. and F.E. Woeste. "Reliability Analysis of Fire Exposed Light-Frame Wood Floor Assemblies," Research Paper FPL 386, U.S. Department of Agriculture, Forest Service Forest Products Laboratory, Madison, Wis., 1981.

10. American Forest & Paper Association. *National Design Specification for Wood Construction,* American Forest & Paper Association, Washington D.C. 1991.

11. ASCE 7-93, *Minimum Design Loads for Buildings and Other Structures,* American Society of Civil Engineers, New York, 1993.

Commentary for 4: Standard Calculation Methods for Determining the Fire Resistance of Masonry

Introduction: Fire resistance ratings are determined by standard test methods and used by the model building codes to address both life safety and property protection. Standards to establish fire resistance are developed and adopted by a consensus group. The code contains mandatory requirements that local jurisdictions adopt as law to govern building design and construction.

Fire resistance is a property of all types of building construction and is related to whether the materials are combustible or noncombustible. Masonry walls are noncombustible materials that have excellent fire resistance characteristics. The fire resistance of masonry walls is determined in accordance with ASTM E119.

There are some instances where a particular type of building construction has not been physically tested in accordance with the ASTM E119 test method. This Chapter provides an analytical method to determine the fire rating for building construction not specifically tested in accordance with ASTM E119. This method is commonly known as the calculated fire resistance.

The resistance of masonry walls to fire is a well-established fact. It is a function of wall mass and thickness. Fire resistance tests have been conducted on walls of solid and hollow units. During the ASTM E119 fire test, the fire resistance of masonry walls is usually established by the temperature rise on the unexposed side of the wall specimen. Few masonry walls have failed due to loading or thermal shock of the hose stream.

The method of calculating fire resistance periods is described in NBS Report BMS 92. The fire technology principles found in Appendix B of BMS 92 were used to develop the procedures described in this Chapter to determine the calculated fire resistance of masonry walls. These include: (1) effects of wall finishes; (2) effects of continuous air spaces; (3) multi-wythe wall construction; and (4) equivalent thickness.

C4.1 Scope

The scope includes assemblies composed of masonry and other components including plaster and drywall finishes and multi-wythe masonry components, which may include other types of inorganic masonry units.

Fire resistance is dependent on the quality of materials, design, and construction, in accordance with the code. Design, construction, and material provisions for masonry structures are published in ACI 530/ASCE 5/TMS 402, *Building Code Requirements for Masonry Structures.*

C4.3 Equivalent Thickness

Recent developments in the clay masonry industry have led to the use of the equivalent thickness method for determining the fire resistance rating of hollow clay masonry. It has been accepted practice to determine the fire resistance rating of concrete masonry units based on the type of aggregate used to manufacture the masonry units and the equivalent thickness of solid material in the wall. Equivalent thickness for clay masonry included in Table 4-1 is based on the same principle as concrete masonry units.

The information contained in this section and Table 4-1 is based on NBS Report BMS 92; NBS

Reports BMS 143; "Annual Report of the Fire Rating Committee"; *Standard Fire Tests of Unloaded Hollow Six Inch Brick/Block Panels (filled with Light Weight-Aggregate; Vermiculite); Fire Resistance of Various Masonry Walls;* and "Fire Test Report #83-13."

Example: Determine the equivalent thickness and fire rating for a nominal $8 \times 4 \times 12$ in. ($200 \times 100 \times 300$ mm) hollow clay masonry unit with the coring pattern shown in Figure 4-1.

1. Determine net volume of units

   Gross volume $= (t)(h)(l)$
   $$= (7.625)(3.625)(11.625)$$
   $$= 321.3 \text{ in.}^3$$

2. Determine core area volume

   $$= 2\,[(4.625)(3.625)(2.875)]$$
   $$+ (4.625)(3.625)(0.625) = 106.9 \text{ in.}^3$$

3. Net volume $= 321.3 - 106.9$
   $$= 214.4 \text{ in.}^3$$

4. Percent solid $= 214.4/321.3$
   $$= 0.667, \text{ unit is } 67\% \text{ solid}$$

5. Determine

   $$T_E = V_d/\{L \times H\}$$
   $$= 214.4/(11.625 \times 3.625)$$
   $$= 5.09 \text{ in.}$$

   From Table 4-1, a hollow clay unit, unfilled, with an $T_E = 5.09$ in. attains a 4 hour rating.

   Extensive testing has established the relationship between fire resistance and equivalent solid thickness for concrete masonry walls. The relationship determines fire endurance based on types of aggregates used and the equivalent thickness of the unit. Table 4-1 contains provisions for required equivalent thicknesses of concrete masonry units manufactured with various aggregate types.

   Units manufactured with a combination of aggregate types are covered by Note 2 of Table 4-1, which is based on wall tests of blended aggregate units. The required equivalent thickness in accordance with Note 2 for a specific fire resistance rating is determined by the following equation:

   $$T_r = (T_1 \times V_1) + (T_2 \times V_2) \qquad \text{(C4-1)}$$

   where

   T_r = required equivalent thickness for a specific fire resistance rating of an assembly constructed of units with combined aggregates (in.)

T_1, T_2 = required equivalent thickness for a specific fire resistance rating of a wall constructed of units with aggregate type 1 and 2, respectively (in.)

V_1, V_2 = fractional volume of aggregate type 1 and 2, respectively, used in the manufacture of the unit

Example: The minimum required equivalent thickness of a wall constructed of units made with expanded shale (80% by volume) and calcareous sand (20% by volume) to meet a 3-hour fire resistance rating is:

T_1 for expanded shale (3-hour rating) = 4.4 in. (112 mm)

T_2 for calcareous sand (3-hour rating) = 5.3 in. (135 mm)

$T_r = (4.4 \times 0.80) + (5.3 \times 0.20) = 4.58$ in. (116 mm)

C4.3.3 Air Spaces or Cells Filled with Loose Fill Material

When the cores of hollow masonry units or air spaces are solidly filled with grout or loose material as indicated, the equivalent thickness of the masonry is considered to be the actual thickness of the masonry.

C4.4.1 Walls with Gypsum Wallboard or Plaster Finishes

The information contained in this Section is based on the Supplement to the National Building Code of Canada, 1990 Edition.

C4.4.1.1 Calculation for Non–Fire-Exposed Side: The fire resistance of masonry walls is generally determined by temperature rise on the unexposed surface, i.e., the "heat transmission" end point. The time required to reach the heat transmission end point (fire endurance rating) is primarily dependent on the thickness of the masonry and the type of aggregate used to make the masonry. When additional finishes are applied to the unexposed side of the wall, the time required to reach the heat transmission end point is delayed and the fire resistance rating of the wall is thus increased. The increase in rating contributed by the finish can be determined by considering the finish as adding to the thickness of masonry. However, since the finish material and masonry may have different insulating properties, the actual thickness of finish may be adjusted to an equivalent thickness of the type of aggregate used in the masonry. The correction is made by multiplying the actual finish thickness by the factor determined from Table 4-2 and then adding the adjusted thickness to the thickness of the masonry. This equivalent thickness is then applied to determine the fire resistance rating from Table 4-1.

C4.4.1.2 Calculation for Fire-Exposed Side: When finishes are added to the fire-exposed side of a masonry wall, the finish's contribution to the total fire resistance rating is based primarily on its ability to remain in place during a fire, thus affording protection to the masonry wall. Table 4-3 lists the times that have been assigned to the finishes on the fire-exposed side of the wall. These "time assigned" values are based on actual fire tests. The "time assigned" values are added to the fire resistance rating of the wall alone or to the rating determined for the wall plus any finish on the unexposed surface.

C4.4.1.3 Assume Each Side of Wall Is Fire-Exposed: Some building codes permit exterior walls with specified set backs from property lines to be unrated from the outside. Thus, in these cases, a calculation assuming the outside of the wall as the fire-exposed side is not necessary.

Where gypsum wallboard or plaster finishes are applied to a masonry wall, the calculated fire resistance rating for the masonry alone should not be less than one-half the required fire resistance rating. This is necessary since the application of additional finishes serves to delay the heat transmission end point without adding any significant contribution to the load-carrying capability of the wall.

Example: An exterior bearing wall of a building is required to have a 2-hour fire resistance rating. The wall will be a concrete masonry unit manufactured with siliceous aggregate. It will be finished on the exterior with 5/8 in. (15.9 mm) of stucco (portland cement-sand plaster) applied directly to the masonry. The interior will be finished with a 1/2-in. (12.7-mm) thickness of gypsum wallboard applied to steel furring members. What is the minimum thickness of masonry required?

First Calculation: Assume the interior to be the fire-exposed side.

1. From Table 4-3 the 1/2 in. (12.7 mm) gypsum wallboard has a "time assigned" of 0.25 hour; therefore, the fire resistance rating that must be developed by the masonry and stucco on the exterior must not be less than 1 3/4 hours (2 hours minus 1/4 hour).
2. From Table 4-2 the multiplying factor for portland cement-sand plaster and siliceous aggregate cmu is 1.00; therefore, the actual thickness of stucco can

be added to the thickness of masonry for use in Table 2-1.
3. Since Table 4-1 does not have required thicknesses for 1 3/4 hour, direct interpolation between the values for 1 1/2 and 2 hour is acceptable. The interpolation results in a required thickness of 3.9 in. of cmu and stucco. Since the stucco is 5/8 in. (19.9 mm) thick, the masonry must be at least 3.27 in. (3.9 − 0.63).

Second Calculation: Assume the exterior to be the fire-exposed side.

1. From Table 4-2 the multiplying factor for gypsum wallboard and siliceous aggregate cmu is 1.25; therefore, the corrected thickness for 1/2 in. (12.7 mm) of gypsum wallboard is 0.63 in. (1.25 × 0.5).
2. Note 1 to Table 4-3 allows 5/8 in. (15.9 mm) of stucco applied directly to the masonry to be added to the actual thickness of masonry rather than establishing a "time assigned" value.
3. Table 4-1 requires 4.2 in. of siliceous aggregate cmu for a 2-hour fire resistance rating. Therefore, the actual thickness of cmu required is 2.94 in. (4.2 − 0.63 − 0.63).
4. Since the thickness of cmu required when assuming the interior side to be the fire-exposed side is greater (i.e., 3.27 in.), this is the minimum cmu thickness allowed to achieve a 2-hr fire resistance rating.
5. Section 4.4.1.4 requires that the cmu alone provide not less than one-half the total required rating. Thus, the cmu must provide at least a 1-hour rating. From Table 4-1 it can be seen that only 2.8 in. of siliceous aggregate cmu is required for 1 hour, whereas 3.27 in. will be provided.

C4.4.1.5 Installation of Finishes: Provisions for finishes on masonry walls are specified in Section 4.4.1.5.1 for gypsum wallboard and gypsum lath and plaster. This criteria establishes the minimum requirements for securing these types of finishes to wood furring, steel furring, or directly to the masonry wall.

Section 4.4.1.5.2 refers to a finish of plaster and stucco directly to masonry walls.

In all instances, these provisions are applicable to the finishes that contribute to the overall fire resistance rating of the assembly. The building code of which this standard has been adopted must also be consulted for applicable provisions that may differ from that contained in this Section. The building code takes precedent over the minimum criteria established in this Section.

C4.4.3 Multi-Wythe Wall Assemblies

In most cases for masonry walls, the fire endurance period will be determined by the temperature rise on the unexposed side of the wall, and its criterion is what the equation is based on. According to the general theory of heat transmission, if walls of the same materials are exposed to a heat source that maintains a constant exposed surface temperature, and the unexposed side is protected against heat loss, the time at which a given temperature will be attained on the unexposed side will vary as the square of the wall thickness.

In the ASTM E119 test, which involves specified conditions of temperature measurement and a fire that increases the temperature at the exposed surface of the wall as the test proceeds, the time required to attain a given temperature rise on the unexposed side will be different from conditions where the temperature on the exposed side remains constant at the initial exposure temperature for any period. It has been found that comparisons fairly consistent with test results can be obtained by assuming the variation to be according to some lower power of n than the second. The fire resistance of the wall can then be expressed by the formula:

$$R = (cV)^n$$

where

R = fire resistance period
c = coefficient depending on the material, design of the wall, and the units of measurement of R and V
V = volume of solid material per unit area of wall surface
n = exponent depending on the rate of increase of temperature at the exposed face of the wall

For walls of a given material and design, it was found that an increase of 50% in volume of solid material per unit area of wall surface resulted in a 100% increase in the fire resistance period. This relation gives a value of 1.7 for n. The lower value of n as compared with 2 for the theoretical condition of constant temperature of the exposed surface is to be expected, as the rising temperature at the exposed surface would tend to shorten the fire resistance period of walls qualifying for higher ratings.

The fire endurance period of a wall may be expressed in terms of the fire endurance periods of the conjoined wythes or laminae of the wall as follows: if R_1, R_2, R_3, etc. = fire resistance periods of walls (or component laminae of walls) having volumes of solid material per unit area of wall surface of V_1, V_2, V_3, etc., respectively, also letting c and n be as defined above,

then for walls in general, $R_1 = (c_1 V_1)^n$, $R_2 = (c_2 V_2)^n$, and $R_3 = (c_3 V_3)^n$.

The fire resistant period of the composite wall will be $R = (cV)^n$

where

$$V = V_1 + V_2 + V_3$$
$$c = (c_1 V_1 + c_2 V_2 + c_3 V_3)/V$$

Therefore:

$$R = (c_1 V_1 + c_2 V_2 + c_3 V_3)^n$$
$$= (R_1^{1/n} + R_2^{1/n} + \ldots R_3^{1/n})^n$$

Substituting 1.7 for n and 0.59 for $1/n$, the general formula becomes:

$$R = (R_1^{0.59} + R_2^{0.59} + \ldots R_3^{0.59})^{1.7}$$

This equation was developed by the National Bureau of Standards in the early 1940s and first appeared in Appendix B of BMS 92. It is noted that the fire resistance period has been expressed in terms of the fire resistance periods of the component laminae of the wall, which need not be of the same material and design.

A composite, multi-wythe wall, i.e., a wall consisting of two or more thicknesses of dissimilar materials, has a greater fire resistance rating than a simple summation of the fire endurance periods of the various layers. The equation developed from BMS 92 permits a calculated fire resistance rating if the fire resistance endurance periods are known for each dissimilar material.

From the same BMS 92 report, it was found that a continuous air space separating wythes of masonry also can increase the fire resistance rating of a masonry wall. A continuous air space can occur between two wythes of masonry or two continuous air spaces can separate three wythes of masonry. By BMS 92, a continuous air space of between 1/2 and 3 1/2 in. (13 and 89 mm) may be estimated by the use of 0.30 and 0.60 for one and two spaces, respectively. The equation provides the formula for calculating the fire resistance ratings involving an air space if the fire resistance periods of the wythes separated by the air space are known. The fire resistance for the various wythes can be obtained from Table 4-1. A partially filled collar joint is not considered an air space. It is virtually impossible for a mason to keep an air space completely clean if it is less than 2 in. (50 mm) wide.

C4.4.4 Multi-Wythe Walls with Dissimilar Materials

This section lists the applicable tables from this document that can be used in conjunction with masonry walls to determine the total fire resistance rating of a multi-wythe wall composed of a combination of concrete, concrete masonry units, or clay masonry units.

C4.4.5 Movement Joints

The fire resistance rating of movement joints is based on fire tests of various joint configurations. Requirements for movement joint materials are covered by the Model Building Codes.

C4.5 Reinforced Masonry Columns

Fire testing of masonry columns evaluates the ability of the column to carry design loads under standard fire test conditions.

Thickness requirements in Table 4-4 for fire resistance ratings of reinforced masonry columns are based on tests of both concrete and masonry columns.

C4.6 Masonry Lintels

Fire testing of masonry beams and lintels evaluates the ability of the member to sustain design loads under standard fire test conditions by measuring the temperature rise of the reinforcing steel.

The minimum cover of 1 1/2 in. (38 mm) required by Table 4-5 is consistent with the code to provide structural and corrosion-resistant protection of reinforcement. Cover requirements in excess of 1 1/2 in. (38 mm) protect the reinforcement from strength degradation due to excessive temperature during the fire exposure period. Cover requirements are provided by masonry units, grout, or mortar.

REFERENCES FOR COMMENTARY 4

1. National Concrete Masonry Association. *A Compilation of Fire Tests on Concrete Masonry Assemblies,* Volume 1 of 5, Summary Report, National Concrete Masonry Association, Herndon, Va., 1993.
2. National Concrete Masonry Association. *A Compilation of Fire Tests on Concrete Masonry Assemblies,* Volume 2 of 5, National Concrete Masonry Association, Herndon, Va., 1991.
3. National Concrete Masonry Association. *A Compilation of Fire Tests on Concrete Masonry Assemblies,* Volume 3 of 5, National Concrete Masonry Association, Herndon, Va., 1991.
4. National Concrete Masonry Association. *A Compilation of Fire Tests on Concrete Masonry Assemblies,* Volume 4 of 5, National Concrete Masonry Association, Herndon, Va., 1991.
5. National Concrete Masonry Association. *A Compilation of Fire Tests on Concrete Masonry Assemblies,* Volume 5 of 5, National Concrete Masonry Association, Herndon, Va., 1991.
6. Brown, P.M. *Western States Clay Products Association—Brick Block Fire Test.* W.R. Grace and Co., Research Division, Washington Research Center, Clarksville, Md., July 1972.
7. National Bureau of Standards. "Fire Endurance and Hose Stream Test of 8-in. Walls of Hollow Brick." NBS Test Report No. TP 1022–22: FP2653, National Bureau of Standards, Washington, D.C., January 1948.
8. National Bureau of Standards. "Fire Endurance of Hollow Brick Walls." *Technical News Bulletin* 35(4), National Bureau of Standards, Washington, D.C., April 1951.
9. Concrete and Masonry Industry Firesafety Committee. *Fire Protection Planning Report No. 13: Analytical Methods of Determining Fire Endurance of Concrete and Masonry Members—Model Code Approved Procedures.* Concrete and Masonry Industry Firesafety Committee, Skokie, Ill.
10. National Bureau of Standards. *Fire-Resistance Classifications of Building Constructions.* NBS BMS 92, National Bureau of Standards, Washington, D.C., October 1942.
11. Structural Research Laboratory. *Fire Resistance of a Brick Cavity Wall System.* Report No. E.S. 6975, Structural Research Laboratory, Richmond Field Station, University of California, Berkeley, Calif., October 1968.
12. National Bureau of Standards. "Fire Resistance of Brick Walls." *Technical News Bulletin,* No. 124, National Bureau of Standards, Washington, D.C., August 1927.
13. National Bureau of Standards. "Fire Resistance of Structural Clay Tile Partitions." NBS BMS 113, National Bureau of Standards, Washington, D.C., October 1948.
14. National Bureau of Standards. "Fire Tests of Brick Walls." NBS BMS 143, National Bureau of Standards, Washington, D.C., November 1954.
15. Fisher and Williamson. "Fire Test Report #83-13." *Two Hour Fire Resistance Test of Higgins Brick Company Solid Grouted 5 Inch Hollow Brick Units.*

16. Foster, H.D. "A Study of The Fire Resistance of Building Materials." *Engineering Experiment Station Bulletin.* No. 104, Ohio State University, Columbus, Ohio, January 1940.

17. "Report of a Standard ASTM Fire Endurance and Hose Stream Test." Building Research Laboratory Report No. T-1748, Ohio State University Engineering Experiment Station, Columbus, Ohio, May 1961.

18. "Report of a Standard ASTM Fire Endurance and Hose Stream Test." Building Research Laboratory Report No. T-1972, Ohio State University Engineering Experiment Station, Columbus, Ohio, March 1962.

19. "Report of a Standard ASTM Fire Endurance and Hose Stream Test of an Unsymmetrical Limited Load Bearing Wall Assembly." Building Research Laboratory Report No. 5477, Ohio State University Engineering Experiment Station, Columbus, Ohio, November 1973.

20. "Report of Standard ASTM Fire Endurance and Hose Stream Tests of Two Non-Loadbearing Unsymmetrical Wall Assemblies." Building Research Laboratory Project 5111, Ohio State University Engineering Experiment Station, Columbus, Ohio, December 1971.

21. "Report of a Standard ASTM Fire Endurance Test and Fire and Hose Stream Test on a Wall Assembly." Building Research Laboratory Report No. T-3660, Ohio State University Engineering Experiment Station, Columbus, Ohio, October 1966.

22. "Report of a Standard Fire Endurance and Hose Stream Test." Building Research Laboratory Report No. T-1971, Ohio State University Engineering Experiment Station, Columbus, Ohio, March 1962.

23. "SCR Brick-Wall Fire Resistance Test." Ohio State University Engineering Experiment Station, Research Report No. 2. Structural Clay Products Research Foundation, Chicago, Ill., September 1952.

24. "Standard ASTM Fire Endurance and Hose Stream Test." Project No. T-1172, Ohio State University Engineering Experiment Station, Columbus, Ohio, November 1959.

25. Structural Engineers Association of Southern California. "Annual Report of the Fire Rating Committee." 1962.

26. *Technical Notes on Brick Construction 16B.* "Calculated Fire Resistance." Brick Institute of America, Reston, Va., June 1991.

27. Technical Notes on Brick Construction 16 Revised. "Fire Resistance." Brick Institute of America, Reston, Va., May 1987.

28. "The Fire Resistance of Brick Walls Made from Clay or Shale." *NBS Letter Circular 228*, National Bureau of Standards, Washington, D.C., June 1927.

29. Troxell, G.E. *Fire Resistance of Various Masonry Walls.* University of California at Berkeley, Calif., December 1967.

30. Williamson, R.B. *Standard Fire Tests of Unloaded Hollow Six Inch Brick/Block Panels (filled with Light Weight Aggregate; Vermiculite).*

31. Ingberg, S.H. and Foster, H.D. Fire Resistance of Hollow Loadbearing Wall Tile. NBS Research Paper No. 37, National Bureau of Standards, Washington, D.C., 1928.

32. Johnson, P. and Plummer, H.C. *Fire Resistance of Structural Facing Tile.* Structural Clay Products Institute, Washington, D.C., August 1948.

33. Plummer, H.C. *Brick and Tile Engineering—Fire Resistance,* Structural Clay Products Institute, Washington, D.C., November 1950, pp. 141–152.

34. ACI 530/ASCE5/TMS 402, Building Code Requirements for Masonry Structures—1995 Edition.

Commentary for 5: Standard Methods for Determining the Fire Resistance of Structural Steel Construction

C5.1 Scope

This Section defines calculation procedures that have, in general, been experimentally validated for specific fire protection materials/systems for structural steel. In most cases, these calculation methods involve the interpolation/extrapolation of standard ASTM E119[1] fire test results. As a result, limitations on the use of these procedures have been carefully established. Additional background is included in a related series of publications developed by the American Iron and Steel Institute.[2–4]

C5.2 Structural Steel Columns

Numerous theoretical and experimental investigations have confirmed that the fire resistance of structural steel columns is a direct function of the mass of the column and the surface area directly exposed to the fire environment.[6,7] As a result, the specified calculation procedures include these variables in the form of weight-to-heated-perimeter (*W/D*) ratios, as defined in this Section and illustrated in Figure 5-1. 28.

C5.2.1 Gypsum Wallboard

Based on a series of fire tests conducted by Underwriters Laboratories, Inc. (UL) and the National Research Council of Canada, an analytical expression has been developed to determine the fire resistance of structural steel columns protected with Type X gypsum wallboard.[8] In order to account for the presence of chemically combined moisture in gypsum wallboard, this equation includes the term W', which is weight per linear foot of the steel column and gypsum wallboard. Since the fire integrity of gypsum wallboard systems is a significant function of the support and attachment methods, the use of this equation is limited to two specific installation methods (Figures 5-2 and 5-3). These installation methods are directly based on UL Designs X526 and X528.[8] Furthermore, since the largest column that has been tested with gypsum wallboard protection is a W14X233, the application of this equation is specifically limited to columns with W/D ratios of less than or equal to 3.65. For larger columns, the thickness of wallboard required for a W14X233 shape should be used.

C5.2.2 Spray-Applied Materials (Wide-Flange Columns)

This Section defines a general equation for determining the fire resistance of wide-flange structural steel columns protected with spray-applied materials. Since the vast majority of these materials are proprietary, this equation includes two constants that must be determined for specific materials based on standard fire test results.

The database should include at least two tests for each of two different column sizes. For the smaller of the two columns, one of the test specimens shall be protected so as to develop the minimum desired fire resistance rating, and the second specimen shall be protected with the maximum intended thickness of fire protection material. For the larger of the two columns, one of the test specimens shall be protected with the minimum intended thickness of fire protection material, and the second specimen shall be protected so as to develop the maximum desired fire resistance rating. These four tests establish the limits governing the use of the resulting equation. These limits include the minimum and maximum permitted thicknesses of protection, the minimum and maximum fire resistance ratings, and the minimum and maximum column sizes. Additional tests may be conducted to modify any of these limits, and these additional tests may involve different column sizes. The material-dependent constants are determined based on all applicable test data using a linear, least-squares, curve-fitting technique or similar statistical analysis.

For some fire protection materials, designs in the UL Fire Resistance Directory include a form of this equation with specific constants and appropriate application limits.[9]

C5.2.2 Spray-Applied Materials (Pipe and Tubular Columns)

Especially for smaller shapes, the fire resistance of pipe and tubular columns protected with spray-applied materials will be somewhat less than the fire resistance of a wide-flange column with the same weight-to-heated-perimeter ratio (W/D) and thickness of protection. In general, the difference is due to heat transfer principles related to the geometry of the cross sections. As a result, different material-dependent constants are required, and these constants must be developed on the basis of standard fire tests as explained in the preceding section. It should be noted that the distinction between wide-flange and pipe and tubular columns mentioned in this discussion applies only in the case of contour protection. For box profile protection (such as the previously described gypsum wallboard system), pipe, tubular, and wide-flange columns are interchangeable based solely on W/D ratios.

C5.2.3 Concrete-Filled Hollow Steel Columns

The parametric equations for hollow steel columns filled with plain concrete (unreinforced) were developed by Lie and Harmathy based on an analysis of 44 loaded fire resistance tests.[10] The experimental results were evaluated using mathematical models to identify important parameters and to generate design equations. Appropriate application limits were established based on the range of tested columns.

C5.2.4 Concrete or Masonry Protection

The equation for concrete-encased columns was derived directly from an expression developed by Lie and Stringer.[11] In order to use this equation, the ambient temperature thermal conductivity and thermal capacity of the concrete must be known, in addition to density and equilibrium moisture content. Since in many cases this information is not available to the designer, a conservative tabulation of these properties is included which may be used in the absence of specific data. This tabulation was developed from data published by Abrams.[12] For masonry units, this procedure has been modified by including the equivalent thickness concept developed for walls.

C5.3 Structural Steel Beams and Girders

For structural steel beams and girders, the same general principles apply as in the case of columns. In this instance, however, the heated perimeter (D) does not include the top of the top flange, which is shielded from direct fire exposure by floor or roof decks and/or slabs.

C5.3.1 Spray-Applied Materials

To some degree, the thermal characteristics of floor and roof decks and slabs influence the fire resistance of beams and girders. As a result, direct design equations have not yet been developed for beams and girders. A beam substitution equation has, however, been developed by Underwriters Laboratories, Inc. (UL) based on an evaluation of existing test data and basic heat transfer principles.[13] This equation can be used in conjunction with approved fire resistance designs and permits the substitution of different beam and girder shapes, provided that the thickness of spray-applied fire protection material is adjusted as a function of the weight-to-heated-perimeter ratios (*W/D*).

C5.4 Structural Steel Trusses

Existing test furnaces cannot accommodate large trusses. As a result, conservative design approaches based on limiting temperatures are required in order to determine the fire resistance of these assemblies. In order to determine the required thicknesses of spray-applied fire protection materials, the previously described column equation should be used. This approach has been recognized by building codes for many years since, during fire tests, columns are simultaneously exposed on all four sides and the limiting temperatures for columns are lower than for other elements, such as beams and girders. Since the bottom chords and vertical and diagonal truss elements can be simultaneously exposed on all four sides, the heated perimeter is determined in the same fashion as for columns. In those instances in which the top chords will be shielded by floor or roof decks or slabs, the heated perimeter may be determined in the same fashion as for beams. In some instances, for practical reasons, users may wish to specify a single minimum thickness of protection for all truss members based on the largest thickness determined for each of the individual truss members.

REFERENCES FOR COMMENTARY 5

1. American Society for Testing and Materials. *Standard Test Methods for Fire Tests of Building Construction and Materials,* ASTM Designation E119-95a, West Conshohocken, Pa., 1995.

2. American Iron and Steel Institute. *Designing Fire Protection for Steel Columns,* 3rd Edition. Washington, D.C., 1980.

3. American Iron and Steel Institute. *Designing Fire Protection for Steel Trusses,* 2nd Edition. Washington, D.C., 1981.

4. American Iron and Steel Institute. *Designing Fire Protection for Steel Beams,* 1st Edition. Washington, D.C., 1984.

5. Lie, T.T. and Stanzak, W.W. *Fire Resistance of Protected Steel Columns,* American Institute of Steel Construction, *Engineering Journal* 10(3), 1973.

6. Society of Fire Protection Engineers. *Handbook of Fire Protection Engineering,* 2nd Edition, Quincy, Mass., 1995.

7. American Society of Civil Engineers. *Structural Fire Protection,* ASCE Manuals and Reports on Engineering Practice No. 78, New York, 1992.

8. Underwriters Laboratories Inc. *Reports on Steel Columns Protected With Gypsum Wallboard Enclosed in a Sheet Steel Cover,* Projects 71NK 2639 and 76NK 8228, Northbrook, Ill., 1975 and 1977.

9. Underwriters Laboratories Inc. *Fire Resistance Directory,* Northbrook, Ill., 1996.

10. Lie, T.T. and Harmathy, T.Z. *Fire Endurance of Concrete-Protected Steel Columns,* American Concrete Institute, *ACI J.,* January 1974.

11. Lie, T.T. and Stringer, D.C. *Assessment of the Fire Resistance of Steel Hollow Structural Section Columns Filled with Plain Concrete,* National Research Council of Canada, Institute for Research in Construction, Internal Report No. 64, Ottawa, Ontario, 1993.

12. Abrams, M.S. *Behavior of Inorganic Materials in Fire,* American Society for Testing Materials, STP 685 Design of Buildings for Fire Safety, Philadelphia, Pa., 1979.

13. Underwriters Laboratories Inc. *Fire Tests of Loaded Restrained Beams Protected by Cementitious Mixture,* Project 82NK 7962, Northbrook, Ill., 1984.

TABLE X3.1 Construction Classification, Restrained and Unrestrained

I. Wall bearing:	
Single span and simply supported end spans of multiple bays:[a]	
(1) Open-web steel joists or steel beams, supporting concrete slab, precast units, or metal decking	unrestrained
(2) Concrete slabs, precast units, or metal decking	unrestrained
Interior spans of multiple bays:	
(1) Open-web steel joists, steel beams or metal decking, supporting continuous concrete slab	restrained
(2) Open-web steel joists or steel beams, supporting precast units or metal decking	unrestrained
(3) Cast-in-place concrete slab systems	restrained
(4) Precast concrete where the potential thermal expansion is resisted by adjacent construction[b]	restrained
II. Steel framing:	
(1) Steel beams welded, riveted, or bolted to the framing members	restrained
(2) All types of cast-in-place floor and roof systems (such as beam-and-slabs, flat slabs, pan joists, and waffle slabs) where the floor or roof system is secured to the framing members	restrained
(3) All types of prefabricated floor or roof systems where the structural members are secured to the framing members and the potential thermal expansion of the floor or roof system is resisted by the framing system or the adjoining floor or roof construction[b]	restrained
III. Concrete framing:	
(1) Beams securely fastened to the framing members	restrained
(2) All types of cast-in-place floor or roof systems (such as beam-and-slabs, flat slabs, pan joists, and waffle slabs) where the floor system is cast with the framing members	restrained
(3) Interior and exterior spans of precast systems with cast-in-place joints resulting in restraint equivalent to that which would exist in condition III(1)	restrained
(4) All types of prefabricated floor or roof systems where the structural members are secured to such systems and the potential thermal expansion of the floor or roof systems is resisted by the framing system or the adjoining floor or roof construction[b]	restrained
IV. Wood construction:	
All types	unrestrained

[a] Floor and roof systems can be considered restrained when they are tied into walls with or without tie beams, the walls being designed and detailed to resist thermal thrust from the floor or roof system.

[b] For example, resistance to potential thermal expansion is considered to be achieved when:

(1) Continuous structural concrete topping is used,

(2) The space between the ends of precast units or between the ends of units and the vertical face of supports is filled with concrete or mortar, or

(3) The space between the ends of precast units and the vertical faces of supports, or between the ends of solid or hollow core slab units does not exceed 0.25% of the length for normal weight concrete members or 0.1% of the length for structural lightweight concrete members.

APPENDIX A

The following is an excerpt from the appendixes of the ASTM E119 Standard Fire Test defining restrained and unrestrained conditions of structural assemblies.

X3. GUIDE FOR DETERMINING CONDITIONS OF RESTRAINT FOR FLOOR AND ROOF ASSEMBLIES AND FOR INDIVIDUAL BEAMS

X3.1 The revisions adopted in 1970 have introduced, for the first time in the history of the standard, the concept of fire endurance classifications based on two conditions of support: restrained and unrestrained. As a result, most specimens will be fire tested in such a manner as to derive these two classifications.

X3.2 A restrained condition in fire tests, as used in this method, is one in which expansion at the supports of a load-carrying element resulting from the effects of the fire is resisted by forces external to the element. An unrestrained condition is one in which the load-carrying element is free to expand and rotate at its supports.

X3.3 Some difficulty is recognized in determining the condition of restraint that may be anticipated at elevated temperatures in actual structures. Until a more satisfactory method is developed, this guide recommends that all constructions be temporarily classified as either restrained or unrestrained. This classification will enable the architect, engineer, or building official to correlate the fire endurance classification, based on conditions of restraint, with the construction type under consideration.

X3.4 For the purpose of this guide, restraint in buildings is defined as follows: "Floor and roof assemblies and individual beams in buildings shall be considered restrained when the surrounding or supporting structure is capable of resisting substantial thermal expansion throughout the range of anticipated elevated temperatures. Construction not complying with this definition

are assumed to be free to rotate and expand and therefore shall be considered as unrestrained."

X3.5 This definition requires the exercise of engineering judgment to determine what constitutes restraint to "substantial thermal expansion." Resistance may be provided by the lateral stiffness of supports for floor and roof assemblies and intermediate beams forming part of the assembly. In order to develop restraint, connections must adequately transfer thermal thrusts to such supports. The rigidity of adjoining panels or structures should be considered in assessing the capability of a structure to resist thermal expansion. Continuity, such as that occurring in beams acting continuously over more than two supports, will induce rotational restraint which will usually add to the fire resistance of structural members.

X3.6 In Table X3-1 only the common types of constructions are listed. Having these examples in mind as well as the philosophy expressed in the preamble, the user should be able to rationalize the less common types of construction.

X3.7 Committee E-5 considers the foregoing methods of establishing the presence or absence of restraint according to type and detail of construction to be a temporary expedient, necessary to the initiation of dual free endurance classifications. It is anticipated that methods for realistically predetermining the degree of restraint applicable to a particular fire endurance classification will be developed in the near future.

INDEX